40 DIY Science Gizmos

For two expert do-it-yourselfers –
my champion speed-knitting mum
and my master craftsman dad – *Ruben*

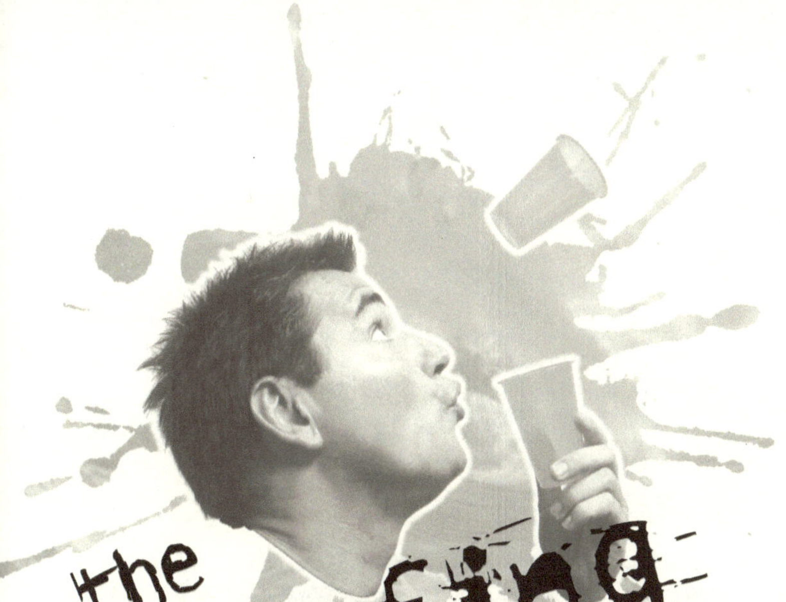

the Surfing Scientist #2

40 DIY Science Gizmos

Ruben Meerman

Published by ABC Books for the
AUSTRALIAN BROADCASTING CORPORATION
GPO Box 9994 Sydney NSW 2001

First published August 2008

National Library of Australia
Cataloguing-in-Publication entry:

Meerman, Ruben.

The surfing scientist: 40 DIY science gizmos / Ruben Meerman.

ISBN: 978 07333 2383 6 (pbk.)

Includes index.

For ages 8 and over.

Science – Experiments – Juvenile literature.

Scientific recreations – Juvenile literature.

Australian Broadcasting Corporation.

500

Internal and cover design by Tou-Can Design

Contents

Introduction

Flashy new toys are great, but there's nothing quite like making your own. Adults get the same satisfaction from do-it-yourself home improvement projects. In this book, you'll learn how to make forty cool gizmos and gadgets that spin, whirl, fly, glide, float, sink, fizz, jump and pop. All you need is a little know-how, a bit of patience and some basic household materials.

Others call this 'playing', but scientists and inventors call it learning. Ask any scientist and they'll tell you that being playful, curious and imaginative is the key to success. Take Sir Christopher Cockerell, who invented the hovercraft, for example. He tested his ideas by playing with coffee jars and cat food tins in the kitchen! Then there's Dimitri Mendeleyev, who invented the Periodic Table of Chemical Elements. This world-famous chart revolutionised the science of chemistry. Dimitri's inspiration came from playing a simple card game called Solitaire.

Clearly, playfulness, creativity and a good imagination are very important in science and kids have heaps of these!

The forty gadgets and gizmos in this book are not my original ideas. Some have been entertaining kids for hundreds of years. Take the Cartesian Diver on page 62, for example. It's named after the famous 17th Century philosopher and mathematician called René Descartes.

Most of the gizmos were invented more recently. The Mexican Jumping Bean on page 18 requires some aluminium foil. This amazing material didn't exist a hundred years ago. My brother taught me how to make these jumping beans when I was about eight. The CD hovercraft on page 56 uses a plastic sports drink nozzle and a CD. Neither of those even existed when I was kid!

The average home is packed with materials for making gizmos and gadgets. There's cling wrap and aluminium foil in the drawers. There are paper towels on the bench, straws in the cupboard and there's a whole fruit bowl full of cool chemicals. There are bread clips, paper clips, bottles, corks, jars, spoons, forks, screws, nails and balloons. There's string, detergent, food dye, rice, vinegar, baking soda and vegetable oil – you get the picture! With all this at your fingertips, you'll soon wonder how you could ever be bored.

I hope you enjoy making the gizmos and gadgets in this book as much as I did. Each one will entertain, amuse and surprise you. They might even inspire you to pursue a career in science some day. Nobody knows where your journey will lead or what you'll achieve, but one thing is certain. If you keep your playful curiosity and imagination alive, anything is possible. Above all, though, I hope these gizmos make you smile. Have fun and happy experimenting!

Oh yeah – and happy surfing, too!

1 Mexican Jumping Beans

These creepy little critters flip-flop about. Use an Easter egg wrapper or colour your foil with a permanent marker for a fun looking bean. And keep your beans safe in an empty matchbox – the kind that had big barbecue matches in it is best.

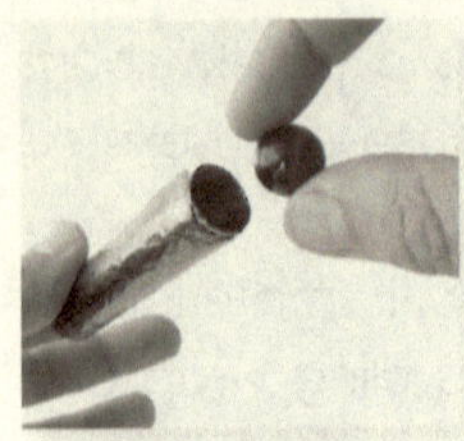

Cut a piece of aluminium foil about 7cm by 12cm. Roll it around a finger or marker pen to make a 7cm long tube. Pop a marble inside the tube.

Squash the ends of the tube shut. Your bean will look like the top photo at this stage. After the next step, it will look like the bottom photo.

Put your bean in a square lunch box. Shake the box so the bean bashes into the sides. Most ice cream tubs' won't do because the corners are too round.

After a bit of shaking, your bean looks like this. Gently roll it around inside the box to make it move. Weird. They're quite delicate, so be gentle. Now go forth and freak out your friends!

What's going on?

The amazing properties of aluminium make these ingenious little critters possible. Aluminium is a light, yet very strong, metal. It's used in many forms to make all kinds of things including planes, boats, soft drink cans, cars and window frames.

Because aluminium foil is thin, it's very malleable (easy to deform). Each impact between the marble and the lunch box shapes your aluminium tube into a capsule with smooth round ends. It's an inside-out form of panel-beating.

When the bean is nice and smooth, the marble can roll from one end of the tube to the other. Because the aluminium foil is much lighter, it flips over when the marble rolls to one end. Then the marble can keep rolling to the other end before flipping the tube again. The flip-flop motion makes your jumping bean appear to be alive. Eeeek!

Real Mexican Jumping Beans

Real jumping beans are actually the seeds of a Mexican shrub that contain a little grub. These are the larvae of a small moth (*Laspeyresia saltitans*). The female moth lays her eggs on the young flowers. The flowers turn into hard seed capsules which trap the egg inside. They are light brown in colour and roughly the size of an uncooked popcorn kernel. After the egg hatches, the larva begins to feed on the pulp inside the seed. When the larva moves, the seed rolls around. It's not quite 'jumping' but very surprising and amusing to watch. After a few months of 'jumping', the larva undergoes metamorphosis. The adult moth emerges through a tiny trapdoor in the seed. The female moths lay their eggs on another flowering shrub and the amazing cycle starts again.

2 Rockin' Roll-on Doll

You can push them down but these dolls always roll straight back up again! Use the design on Page 92 or design your own. Colour in with your favourite team colours and you've got a cool desktop mascot.

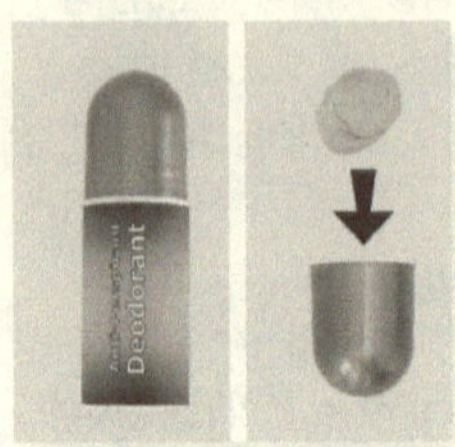

Save the lid from an empty roll-on deodorant bottle. Insert a big blob of adhesive poster putty.

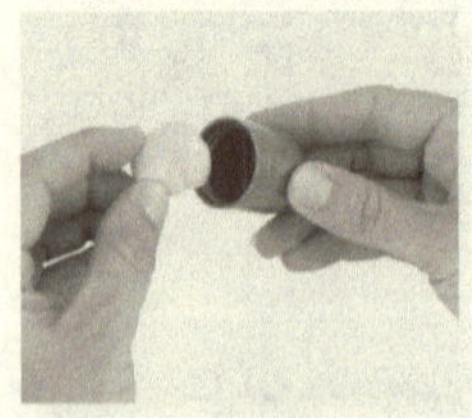

Squash the putty right down into the lid. Test that the lid rolls upright when you lay it on its side. If not, add or remove putty until it does.

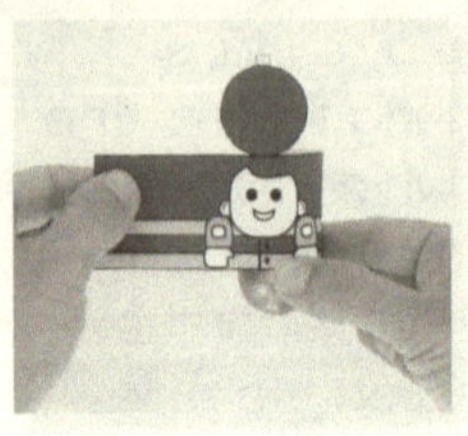

Photocopy or trace the design on Page 92, colour it in and cut it out. Wrap it around the lid and secure with sticky tape. Fold and stick the circle down.

Ta-da! Just look at this happy little snapper-head! He really is a superb little *roll model*... kezam!

SCIENTISTS
patrol these waters

What's going on?

The top part of a roll-on deodorant bottle lid is a shape called a hemisphere (half a sphere). Adding a blob of poster putty makes it bottom heavy, so it always rocks back into the upright position. Scientifically speaking, that's because the putty and lid have a low centre of gravity.

The centre of gravity is a kind of average position for an object's overall mass. If it can, an object's centre of gravity always ends up in the lowest position possible. The smoothness of the hemisphere lets your doll roll back so its centre of gravity is as low as it can get.

Rolling ball technology

Roll-on deodorant bottles have the same design as one of the world's most useful inventions – the ballpoint pen! Before it was invented, people used a quill (bird's feather), which they sharpened and dipped into ink. It worked, but smudges were almost inevitable. The metal fountain pen replaced the quill, but it smudged easily too and worse still, tore through paper if you pressed too hard.

The first patent for a rolling ballpoint pen was filed in 1888 by John L Loud. The design had too many problems to be successful, but the idea was brilliant. Other inventors started experimenting with different designs, but most of the pens had big problems. It was a Hungarian newspaper editor called László Bíró who finally made the idea work. Driven to frustration by fountain pens, László and his brother George, who was the chemist, solved the problems. They successfully patented their design in 1938. It was so popular that 'Biro' became the generic name for all ballpoint pens.

3 Acrobatic Cutlery

Here's a classic science trick with a fork, a spoon and a toothpick. Make sure you ask for an old metal fork and spoon to use, though. You'll be in big trouble if you bend the best ones in the drawer!

Carefully push a spoon between the prongs of a fork. Slightly bend the two inner prongs to one side of the spoon's bowl. Bend the outer prongs on the other side.

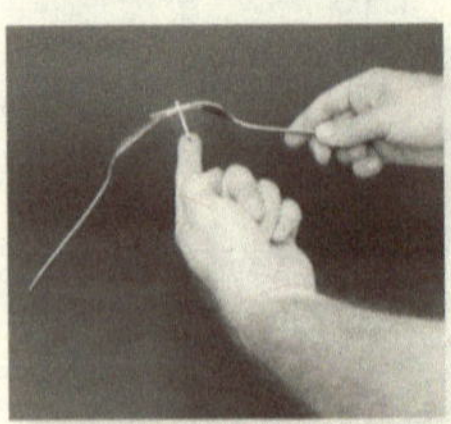

Push a toothpick in between the two inner prongs. You should now be able to 'balance' your cutlery on the tip of your finger.

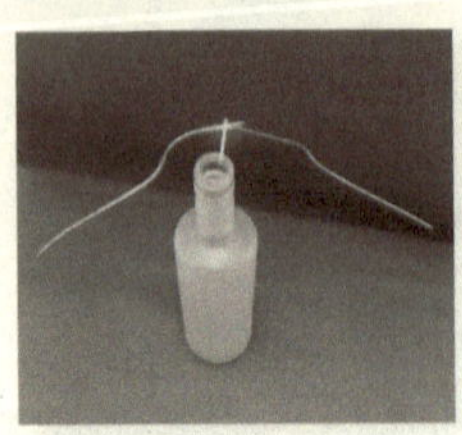

Carefully balance the fork and spoon on the toothpick. Do this on the rim of a bottle or any object you like (be careful and wear shoes in case they fall on your toes!)

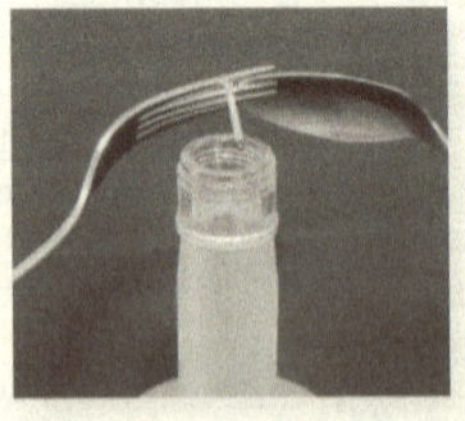

It looks very unstable but the fork and spoon will rest easily on the rim like this. It's a surprisingly neat little balancing act.

Sandologist ON DUTY

What's going on?

The arrangement you made looks like it's balancing on the bottle, but it's actually hanging. The difference between hanging and balancing is usually quite obvious, but this set up gives a false impression.

Whether an object is balancing or hanging depends on where its centre of gravity is, relative to its points of support. The centre of gravity of a body is the average position for all its mass. For a marble, it's right in its centre. But for a broom, the centre of gravity is closer to the brush, not half way up the handle. That's because the mass of a whole broom is not distributed symmetrically. An object is balancing if its centre of gravity is above the point (or points) it is balancing on. Hanging is when an object's centre of mass is below its point of support.

Now look at where the heaviest parts of the forkand spoon are. Most of their weight is in the handles so their combined centre of gravity is actually belowthe toothpick. They are actually hanging, not balancing... but don't tell anyone!

Fork etiquette

Forks didn't come into popular use until the early 1600s. Before then, they were actually considered a bit vulgar. The prongs are called tines. In our culture, it's considered good etiquette to keep the fork in the left hand with the tines pointed down. The handle should be under your palm. This makes eating peas very tricky!

In the United States, however, the correct etiquette is different. There, it's fork in the left hand with tines up but only for cutting. Then it's knife down, switch fork to right hand, put food in mouth, fork down, chew and swallow. Repeat for next bite. No, I'm not kidding! It's all a bit confusing but like, um – whatever. Pass me the chopsticks, please!

4 Tightrope Walker

You'll need a straw, a single-groove screw and four blobs of adhesive poster putty – one for the body, one for the head, and two for dumbbells. Get it 'balanced' just right, and you've got a groovy tightrope walker!

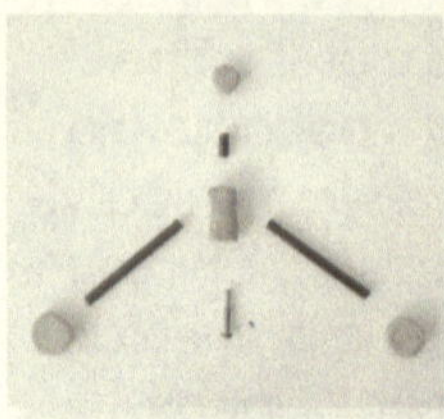

Cut two equal lengths of straw for the arms and a short piece for a neck. Make the body roughly square and the head and dumbbells round.

Press the straws and screw firmly into the adhesive putty like this. Test whether it 'balances' on the end of your finger. Be patient and make only tiny adjustments to get it just right.

Set up a fishing line or cotton thread tightrope between two tall objects (I used two empty bottles filled with coloured water for mine).

If you raise one of the bottles, the tightrope walker will glide down without falling off. Not bad for a kooky little puppet, eh?

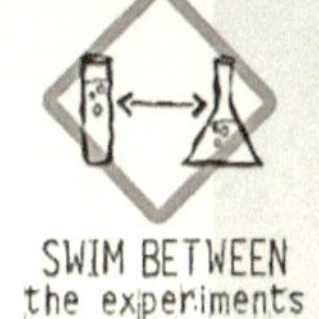

What's going on?

This little puppet looks like it's balancing, but just like the Acrobatic Cutlery, it's not. The combined centre of gravity of all the parts is actually below the head of the screw, so it's really just hanging.

For your tightrope walker to 'walk' on the string, its centre of mass must be slightly below the bottom of the screw. If you're struggling to make this work, take it slowly and only make tiny adjustments at any one time. A well balanced tightrope walker should glide along the cotton or line without falling off.

Real Tightrope Walking

Now that you know this gizmo's secret, you might think real tightrope walking isn't such a big deal. But take a closer look at these amazingly brave folks and you'll notice they really are balancing. Yikes! Some tightrope walkers use a long bar for stability, but this doesn't lower their centre of gravity below the wire, so it's not really cheating! 19th century tightrope walker William Hunt, also known as the Great Farini, could do a headstand on a wire! He amazed visitors to Niagara Falls in North America by walking across a tightrope stretched over the falls. Just about everything imaginable has been done on tightropes – from bike riding to juggling. Another outstanding skywalker called Charles Blondin crossed Niagara Falls, too. For added flair, he stopped halfway to cook an omelette on a stove he carried onto the rope. Now doing that really is showing off!

By the way, tightrope walking is called funambulism. A tightrope walker is a funambulist. Funambulism is derived from the Latin word funambulus, which means rope-dancer. Sounds funny, but it's true!

5 Paper Bugs

Boring dinner parties spring to life with shrieks of terror when a Paper Bug scurries across the table. So cheap, so simple and yet so much fun!

Grab a lemon and a paper towel or napkin. Other small round fruits and vegetables work, too.

Twist the corners of the paper tight to make four freaky 'legs'.

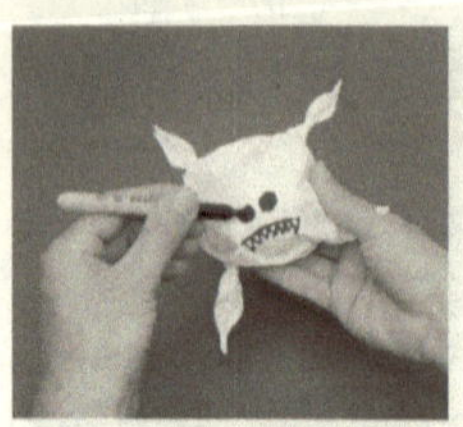

Pop the lemon under the paper bug. Use a marker pen to draw some eyes and a scary mouth with fangs for extra freakiness.

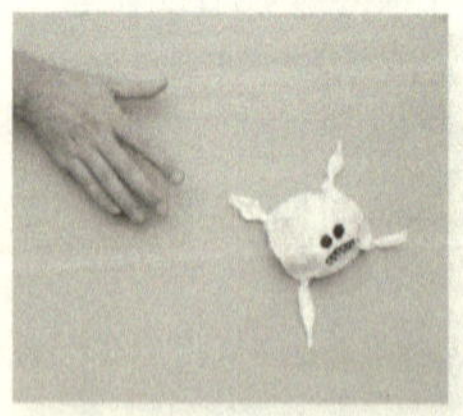

Gently push your bug along a smooth table so the lemon rolls. It will scurry all over the place, terrifying your victims as it goes!

SCIENTISTS
patrol these waters

What's going on?

A rolling lemon doesn't usually frighten anyone. But disguised as a bug under a paper towel, it will wreak havoc!

The freakiness is mainly due to the unexpected motion of the rolling lemon inside. Any round but not perfectly spherical object will roll in a slightly unpredictable manner. When you can see the rolling lemon, its motion isn't surprising or freaky at all. But nobody expects a scrunched-up paper towel to move this way. With the lemon hidden from view, its motion resembles that of a scurrying insect or small animal.

Fears vs phobias

Anything that moves unexpectedly, or in an unexpected way, can trigger a fear response. Fear is a healthy and very normal reaction. People are naturally nervous of insects or spiders, because some can sting or bite. But for most people, a small and sudden shock soon passes. Small controlled frights can even be quite fun – that's why we love scary films so much!

An abnormally acute or persistent fear is different. These are called phobias (from the Greek word phobos). A fear of insects is called 'entomophobia' and a fear of spiders is called 'arachnophobia'. 'Ornithophobia' is a fear of birds and 'amathophobia' is a fear of dust. 'Octophobia' is a fear of the number eight and 'xenophobia' is a fear of strangers.' Acrophobia' is an abnormally severe fear of heights.

For people who suffer phobias, an encounter with the object of their fear can be terrifying and is never fun or funny. It can stop them participating in many of the activities the rest of us take for granted. Luckily, most phobias can be cured with controlled exposure to the problem animal or situation.

6 Paper Spiral Spinner

Bored? Spent all your pocket money? No worries. This spinning gizmo is easy to make, costs nothing and looks great. Ask an adult to supervise or light the candle for you. Then it'll spin like a charm and voila! Happy days are here again!

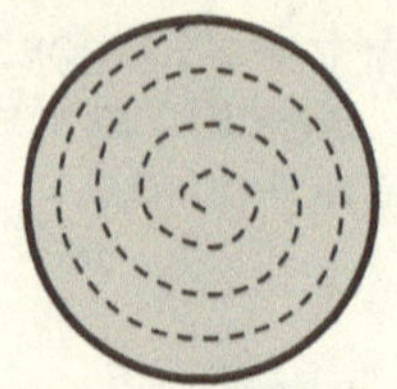

Draw a circle on a piece of coloured paper by tracing around a CD. Draw a spiral inside the circle like this. You can use white paper and add a coloured pattern first for an even cooler spinner!

Cut the circle out carefully. Now cut along the spiral you drew. Attach a long piece of string to the centre of the spiral with sticky tape.

Tie the other end of the string to an object so the spiral hangs at least 20cm above a candle. Ask an adult to light the candle.

The spiral will spin as long as the candle is lit. Groovy!

Safety note: never leave a burning candle unattended!

What's going on?

The candle flame produces heat. Hot air rises, so there is an invisible stream of rising air directly above the candle. If you could see it, you would notice that the rising air gets deflected sideways by the spiral. Paper is very light, so this deflection causes your spiral to spin around and around. Depending on how you drew the spiral, your spinner will rotate clockwise or anti-clockwise.

World's Most Important Invention?

It's difficult to rate an invention's importance, but paper would have to be very high on the list. Business, governments, literature and toilets would be very different without the stuff. The papermaking process is considered one of the four great inventions of ancient China.

Paper is made from cellulose fibres, which are derived mainly from wood. Wood is strong because the cellulose fibres are held together by a substance called lignin. To remove the lignin, wood has to be pulped. This takes an extraordinary amount of energy (try breaking a plank of wood in half and you'll see why). To make white paper, the wood pulp has to be bleached with powerful chemicals. Some of these chemicals are very harmful to the environment. That's one of the reasons we should use paper wisely and try to minimise waste.

The cellulose fibres are tightly interwoven which keeps them bound together. Starch and chemicals are added to office paper for more strength. The bleached pulp is rolled into sheets, flattened and dried. Then it is coated with china clay and polished. It takes a lot of work to make a piece of paper!

7 Magic Flowers

These cheery little flowers are guaranteed to make you smile. Photocopy or trace the design on Page 92 or design your own. Then colour them in, cut them out, fold them up and float on a sauce of water. Smiles all round.

Trace or photocopy the flower on Page 92. Colour in with crayons or pencils (felt pens will run). Then neatly cut your flower out with scissors.

Fold the petals over one by one so the coloured side is hidden. Put a teaspoon of water in the centre of a saucer and float the flower on top.

Immediately, the petals will start to slowly open like this.

In about thirty seconds, your flower has bloomed. Make a magic flower for someone who's feeling sad and I guarantee you'll make them smile.

BEACH CLOSED

serious experimentation underway

What's going on?

When dry paper absorbs water, it expands slightly. This is why the creased petals slowly open up when they get wet. It's also why office paper gets wrinkles if you spill water on it. When you float your flower in water, the petals open up in the order you folded them. Try designing flowers with longer or wider petals and folding them in different orders. But please use paper wisely and draw as many flowers as you can fit on a single sheet to minimise waste.

What's one sheet of paper?

If you think recycling your paper doesn't make much difference, grab a calculator and be surprised.

First, we need some facts. At the beginning of 2008, the Australian population was about 21 million and growing. One sheet of A4 office paper weighs about 5 grams and costs about 1c. It might seem a pretty trivial question, but what if every Australian wasted a single sheet of paper every day? At this rate, we would be wasting147 million sheets per week. Multiply by 52 weeks and that's over 7 billion – 7 644 000 000 – sheets every year. At 1c a sheet, all this wasted paper costs $76 440 000. New cars cost about $30 000 so that's enough to buy 2 500 of them a year!

At 5 grams per sheet all this wasted paper adds up to 38 220 000 kg. An average car weighs about 1 000 kg, so that's the same as 38 220 cars! A fresh pack of photocopier paper containing 500 sheets is about 5cm thick. If you could neatly stack all this wasted paper, it would stand around 760 metres tall. That's higher than the world's tallest skyscraper! So there you have it. Using paper wisely really can make a difference.

8 Robo-Manga Head

Here's a cool variation on a classic old trick. First, stick two plastic cups to a balloon without any glue or sticky tape – cool! Draw on a face and you've got a neat hanging decoration or a snazzy head for your next cardboard robot!

Inflate a balloon to around the size of an orange. Press two plastic cups firmly against the balloon, then continue inflating.

When you've finished inflating, let go of the plastic cups. They should stay right where you were holding them. Cool! Be careful, though – they come off pretty easily.

Use a permanent marker to draw a face. Now you've got a cool head for your next cardboard robot creation. Or, tie on a string and you've got a cool decoration to hang in your room.

I used a Manga drawing tutorial I found on the internet to draw these cute eyes. Manga is a popular form of Japanese animation. Pretty cool, eh?

What's going on?

The cups stick to the balloon because they have a partial vacuum inside them. No, I don't mean half a vacuum cleaner. A partial vacuum is a region at lower than atmospheric pressure. Now if you're wondering how on earth that vacuum got in there, try it again. This time, however, watch carefully to see what's going on inside those cups.

When the balloon is small, it dips far into the cup. Because you are pressing the cup onto the balloon, no air can get in or out. When you inflate the balloon, it gets bigger so it doesn't dip into the cup as far. The space inside the cup has increased a bit. That means the air inside the cup now has more room to move around. This is how the vacuum forms. Whenever a sealed volume of air suddenly gets more room to move around, the pressure goes down. The air pressure outside, however, remains constant. It's actually the air outside squashing the cup firmly against the balloon. Hey presto, your Robo-Manga Head's ears are stuck without glue!

Suction cups

Suction cups like the ones on toy darts use the same principle to stick to a smooth surface like glass. By pressing them firmly against a window, the air between the suction cup and the glass gets forced out. When you stop pressing, the cup returns to its original shape. This increases the space between the cup and the glass and that forms a vacuum. The air pressure outside the suction cup squashes the cup firmly against the glass, so it sticks. Friction stops the dart from sliding down. This works so well that you can even use a suction cup with a hook on it to hang things from a window or mirror.

9 O-Wing Glider

You have to see these gliders in action to believe they really can fly. Even then you'll probably pinch yourself to make sure you're not dreaming! They glide gracefully and easily accomplish surprisingly long flights. Get cracking and you'll be amazed!

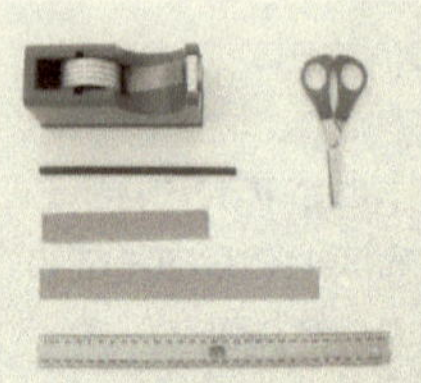

All you need is some stiff card, sticky tape, scissors, a ruler and a straw. Cut two strips of stiff card each 2.5 cm wide. Make one 25cm and the other 15 cm long.

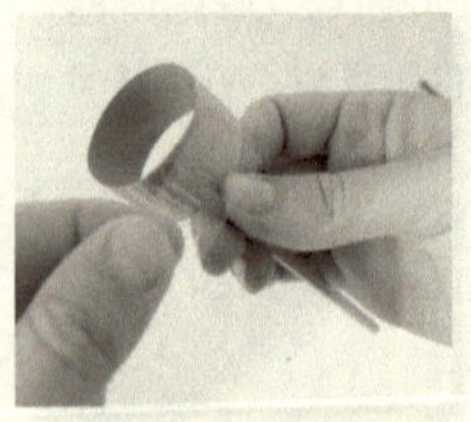

Turn the strips into cylinders and secure the ends (inside and out) with sticky tape. Then stick the cylinders to either end of the straw.

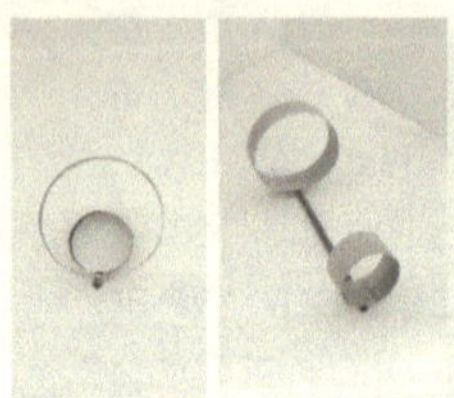

Align the cylinders on the straw like the picture on the left and your O-Wing glider is complete. Easy!

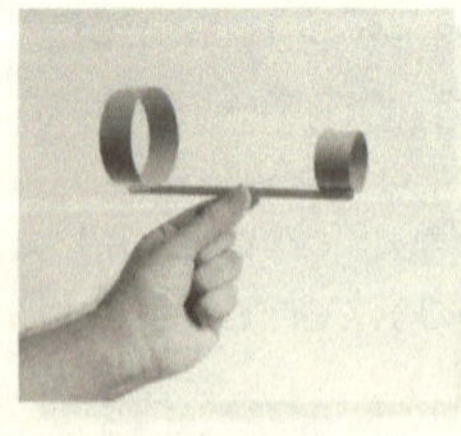

To make it fly, gently throw your O-Wing Glider with the small cylinder at the front. Like all gliders, it will fly best when it's not too windy.

What's going on?

The fact that these bizarre contraptions fly at all is surprising enough. Watching one sail elegantly through the air is nothing short of amazing!

O-Wing aerodynamics is perplexing, because this glider looks nothing like a typical one. But with the wings correctly aligned, it is amazingly stable in flight. Even if a flight starts out wobbly, the straw usually ends up hanging down and the flight continues smoothly.

Aligning the wing cylinders correctly on the straw is the key to a stable O-Wing. If yours flies in a corkscrew trajectory, the wings are probably not aligned. It's easy enough to realign the wings, but if it still won't fly, start again. Just don't give up. You won't believe your eyes when you see one fly...through the sky...okay, bye!

X-Planes

Unlike the X-Wing Fighters in 'Star Wars', X-Planes are very real. The 'X' stands for experimental aircraft and many of the X designs are truly bizarre. Some are only conceptual, but many have been built and tested in the air. The Boeing-built, pilot-less X-50A, for instance, looks more like the Jetsons' car than a plane. It has a spinning wing on top and very short fixed wings at the front and back.

One of the weirdest-looking contraptions ever flown was designed, built and tested right here in Australia. Karl Dehn called his gliding creation the 'Ring Wing'. It looked like a flattened donut with a cockpit and a tail. The locals in Benambra, Victoria who saw Karl's Ring Wing tagged it 'The Flying Dunny Seat'. Photos of the Ring Wing can still be found on the internet. Unfortunately, however, a motorised version was never built. That's a shame because seeing a toilet seat fly past would surely brighten up anyone's day!

10 Jumb-O-Wing Gliders

So you've completed your basic O-Wing training. Now it's time for the advanced course. Try experimenting with wing sizes, fuselage length and ballast first. You'll discover that when it comes to O-Wings, the sky really is the limit!

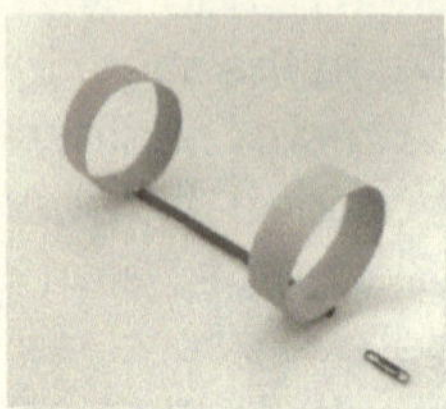

Try experimenting with different wing sizes. You'll find equal sized wings don't fly. But add a paper clip to the front and you're away again.

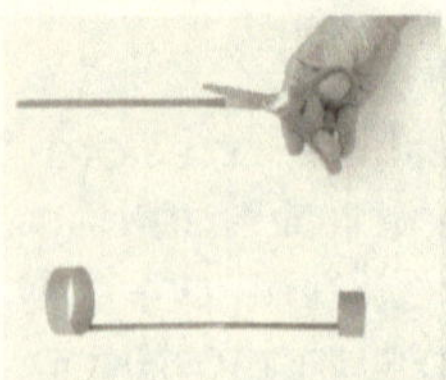

Snip a small slit in the end of a straw. Insert it into a second straw to make a double length fuselage. Will it fly? Yes! But try it and see for yourself.

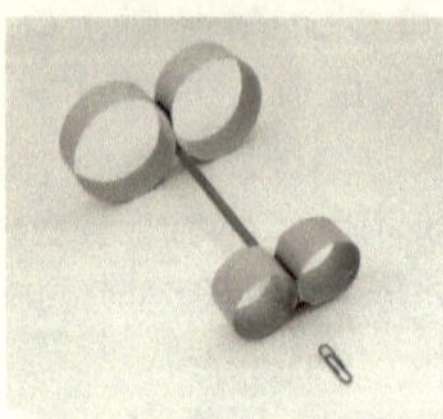

Get creative. Start with two sets of wings on one straw like this. I guess you'd call it an O-O-Wing? Add a paper clip for ballast if it doesn't fly very well.

This Jumb-O-Wing model has 7 rings. It descends a bit faster than the others, but still glides beautifully. I call it the 007 Wing!

Sandologist
ON DUTY

What's going on?

Experimenting with O-Wings is easy, because you can design, make and test them quickly. Try adding little paper fins or wings. The possibilities are endless. You might even stumble onto a new design which revolutionises the science of aviation. This is not some silly cliché to make you feel good. Every scientist or inventor who ever lived was once an inquisitive, imaginative, playful kid. But they all learned an important lesson – there is no such thing as a 'failed' experiment. Even a disappointing outcome teaches you something so you'll do it differently next time!

Nature's Gliders

Long before humans took to the skies, many plants and animals had evolved amazing aerial feats. The Australian sugar glider possum is one such critter. It has a membrane running from its fifth finger all the way to its ankle. Stretching this membrane allows the possum to glide up to fifty metres through the air! Feather-tailed possums use their large tails to steer during a glide. Some species of squirrels and lemurs have awesome gliding skills, too.

Many plants also use cool aerodynamic tricks to disperse their seeds. Sycamore tree seeds have winglets so they spin like tiny helicopters. Dandelion seeds have a fluffy tip that catches the wind so they can drift high into the sky. But the most elegant gliding seeds by far are found in Borneo. The seeds of *Alsomitra macrocarpa*, a liana, or woody climbing plant, glide in beautiful arcs. These seeds are very flat and about the size of an adult's hand. They have two super-thin winglets which allow them to glide away from the parent plant. If they land in a good spot, they will germinate and grow a new liana.

11 Mini Boomerangs

A wooden boomerang is one of the coolest inventions on the planet. But let's face it – it's not real good indoors! The Mini Boomerang solves the problem. It's light, so it won't break a window, and it really does come back!

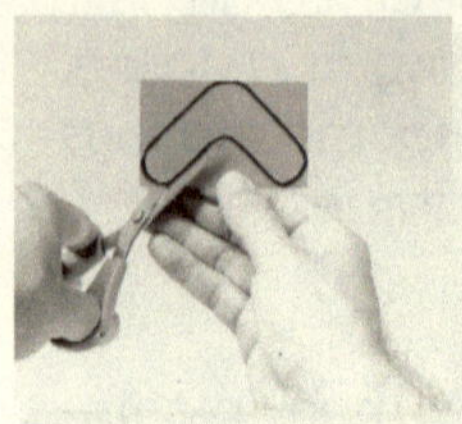

Draw the shape of a boomerang on a business card and cut it out carefully. Make the wings about 2.5cm wide.

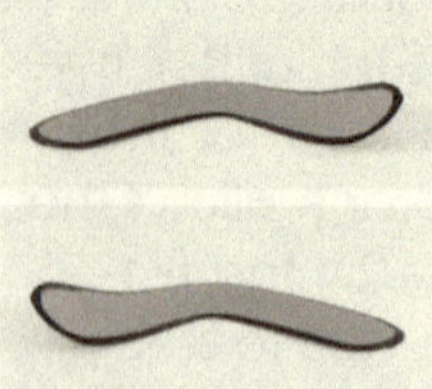

If you are right-handed, bend the right wing up as in the top picture. If you are left- handed, bend up the left wing as in the bottom picture.

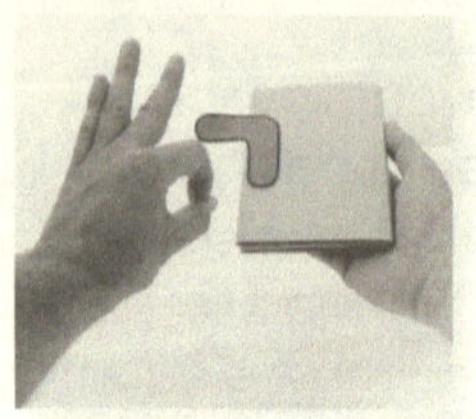

Lay the boomerang on a small book like this. I'm left-handed, so mine is hanging over the left edge.

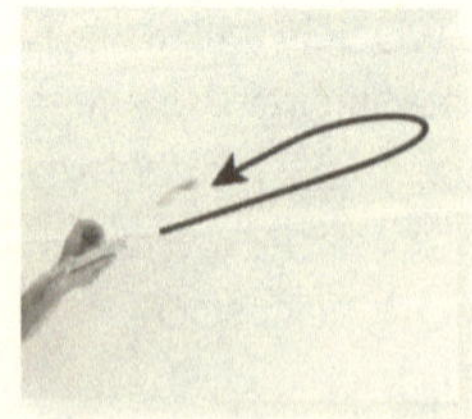

Flick the boomerang as hard as you can, so that it spins fast as it shoots off the book. It will fly up to two metres away and return. Cool!

What's going on?

A boomerang looks deceptively simple, but performs an amazing aerodynamic feat. Throwing it correctly takes skill and practice, but one of the key ingredients is spin. Spinning makes the tips of a boomerang travel at different speeds from the centre. In the diagram below, the boomerang is travelling to the right and spinning clockwise. On the left, the white tip is spinning towards the direction the boomerang is travelling. That means the white tip is actually travelling a bit faster towards the right than the centre. In the next picture, the white tip is spinning away from the direction of the boomerang. That means the white tip is now travelling a little bit slower to the right than the centre of the boomerang. When it rotates, it will be travelling faster again, then slower and so on. So during half a rotation, the white tip is moving faster than the rest of the boomerang. During the other half, it's moving slower.

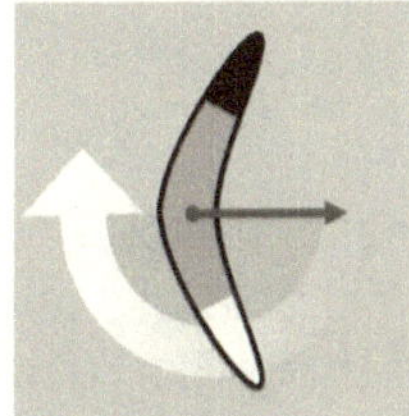

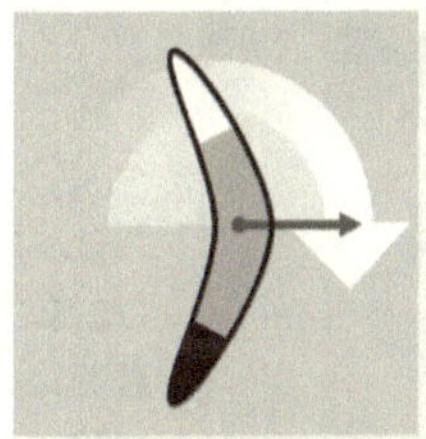

Now the faster a wing moves through air, the more lift it generates. Below a minimum speed, a wing won't generate enough lift. That's why planes need long runways to get off the ground. This is the boomerang's secret. A boomerang's wings move faster while spinning towards the direction of flight. This generates more lift on one side which tilts the boomerang at a slight angle. This tilt causes the boomerang to fly in an arc and return to the thrower. Amazing, huh?

12 Hovercopter

Operating a blow dryer isn't rocket science, but please ask an adult to supervise for safety. The only problem is they'll be so impressed by your Hovercopter that you might not get another turn!

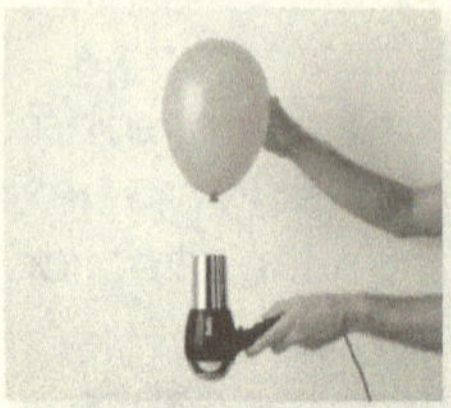

Try this first. Point a blow dryer at the ceiling and switch it on. Hold a balloon in the air stream. What happens when you let go of the balloon?

Amazingly, the balloon doesn't fly off and fall to the ground. Instead, it stays put and hovers in the air stream! Pretty cool, but a bit wobbly.

Now tie or sticky tape a small object, like a coin or bottle lid to a piece of string. Tie your 'cargo' onto the Hovercopter.

Your Hovercopter will be much more stable in the air stream than before. You can even manoeuvre it around obstacles.

SWIM BETWEEN the experiments

What's going on?

A balloon gets stuck in a jet of air because of the Coanda Effect. It's named after Henri Coanda, the Romanian scientist who first described it in the 1930s. Henri noticed that jets of fluids (liquids and gases) tend to stick to curved surfaces. As long as the curve is not too sharp, the jet follows the surface. This is what happens to the invisible jet of air blasting from your blow dryer. It completely envelops the balloon.

But here's another surprising thing about fluids that helps your Hovercopter fly so well. When fluids are flowing over a surface, the pressure on that surface is decreased. The faster the flow, the less pressure on the surface. This is called the Bernoulli Effect. The higher pressure further away from the balloon keeps it stuck firmly in the airstream.

Wind Tunnel Testing

Air is invisible, so studying airflow over, through or around objects is tricky. Engineers use wind tunnels to test the aerodynamic properties of new cars, planes, and even skyscrapers. Scale models in the wind tunnel allow for careful measurements of lift, drag and pressure. Engineers fine tune their designs by watching a stream of smoke or dye flow over models.

Wind tunnels are used to test whether a new skyscraper will stand up to strong winds too. Scale models of entire cities are made to test how a new building will affect the existing structures.

If you enjoy building model cars, planes or buildings, you're actually teaching yourself some valuable skills. Some call this 'playing' – but, in reality, it's learning. So next time dinner's ready before you are, try this excuse: 'One moment, please – I'm just fine-tuning my engineering skills!'

13 Rotocopter

Now that you've earnt your Hovercopter stripes, it's time to crank up the cool factor to the max. The Rotocopter has tiny fins for rotation during flight! You can fly it around obstacles and even land it in jars. Now that is super cool!

Wrap a piece of string around the middle of a balloon to act as a guide. Staying close to the string, use a permanent marker to draw a line right round the balloon.

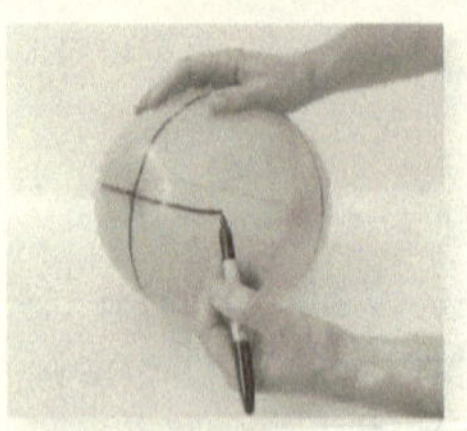

Next, wrap the string over the balloon from top to bottom. Follow the line with your marker again. Repeat this step to draw another line that crosses the one you just drew.

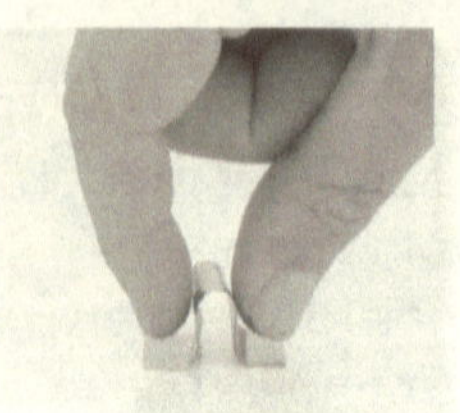

Take a piece of sticky tape 5 cm long and bend it into a loop like the one in this picture. Squeeze the loop together to make a tiny rotor blade. Make a total of four rotor blades.

Stick the rotor blades to the side of the balloon where the lines cross. Set each one at a 45 degree angle as in this picture. Repeat this for a total of four rotor blades.

SCIENTISTS
patrol these waters

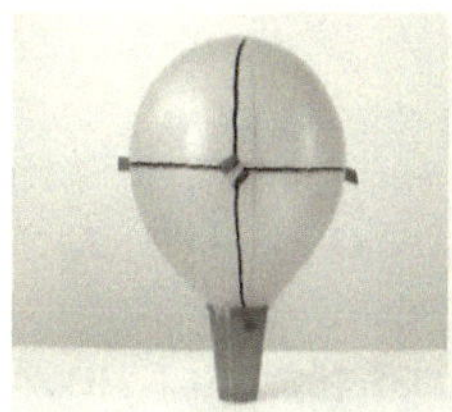

Ready for blast off! Aim a blow dryer at it and your Rotocopter takes to the air. The rotor blades make it rotate so it looks like a cool spinning UFO.

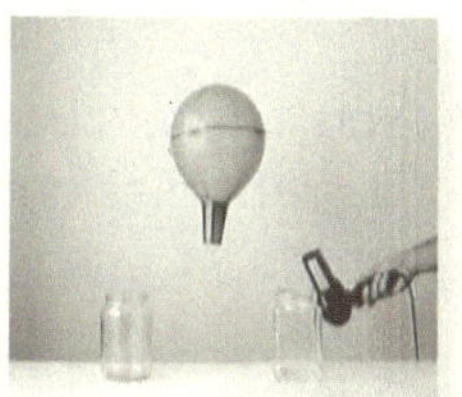

Rotocopter Obstacle Course

Check out the flight sequence above. I set up two large jars. The first photo is pre- take-off. I took the middle photo mid-flight and the final one during landing. Use more jars to set up a Rotocopter Obstacle Course. You have to take off and then land the Rotocopter in each jar. The fastest time taken to reach the last jar wins.

Windmill Power

Without rotor blades, a Rotocopter is just another Hovercopter – still cool, but no spin. Wind pushing against the rotor blades of windmills work the same way. The race for environmentally friendly electricity has greatly improved windpower technology in the past 15 years. In fact, wind farms have been cropping up all over the place. That's great news for the environment! But long before electricity was lighting up homes, Holland (where I was born) was covered in more primitive windmills. Instead of generating electricity, Dutch windmills turned huge stone wheels to crush wheat into fine flour. And those Dutchies need lots and lots of flour because, boy, do they love their pancakes!

14 Tornado Bottle

Water swirling down the sink or bath looks just like a tornado, but doing it just for fun is a waste of water. This gizmo solves the problem. You'll get a better view of your 'tornado' without wasting a drop.

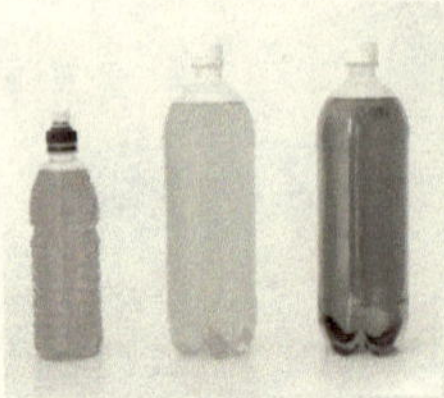

You'll need two large empty soft drink bottles and a sports drink nozzle.

Remove the sliding part of the sports drink nozzle. This is pretty hard to do, so you'll need an adult helper.

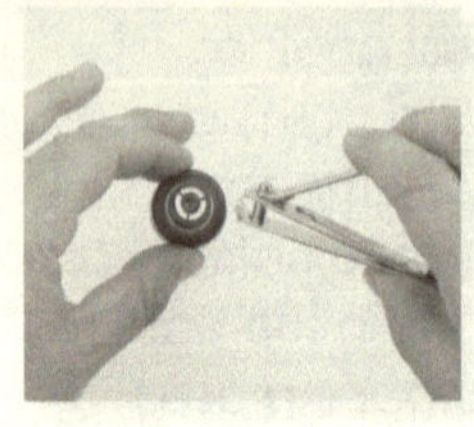

Use nail clippers to cut the stems that are holding the centre piece in place.

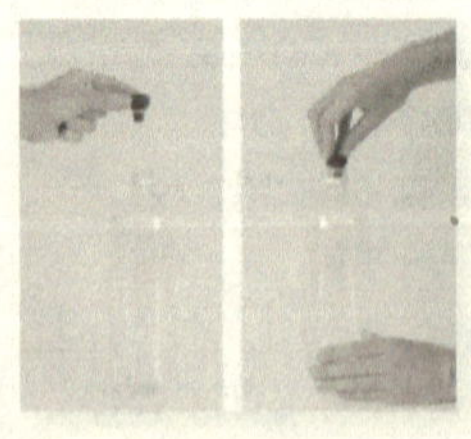

Try to remove as much of the stems possible. Discard the centre piece. One step to go!

Sandologist ON DUTY

Fill one of your bottles to the brim with water and screw on the nozzle. Position the empty bottle on top. Use strong, stretchy, waterproof tape to hold the bottles together.

Hint: Stretch the tape as you wind it onto the bottles

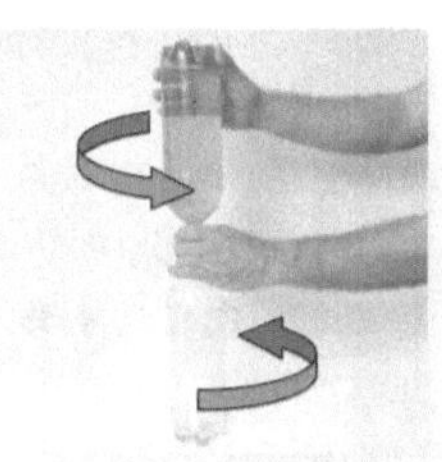

To make a tornado, hold the top bottle and wiggle it in small circles. This gets the water swirling fast. You can stop wiggling once the tornado forms.

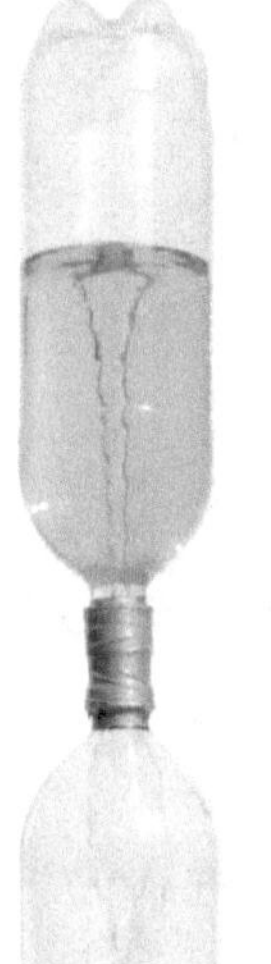

Hint: Don't be too rough. You don't want the bottles to come apart!

What's going on?

Real tornadoes happen in air, not water. Your bottled 'tornado' may not be as terrifying as the real thing, but it works on the same principle. When the water is in the top bottle, gravity pulls it down. But for water to pour down into the bottom bottle, an equal volume of air has to pour up into the top bottle. If you turn the bottle over very carefully and hold it still, all the water just stays in the top bottle. When you wiggle the bottle, the water starts to swirl. The swirling action keeps the rising air bubbles centred. When enough bubbles are lined up, they join together and a beautiful vortex forms. The long tubular hole in the vortex allows air to pour up while the water pours down.

Real Tornadoes

Real tornadoes form when a huge body of cold air gets blown over the top of warm air. Cold air is denser than warm air, so it descends. The cold air in a tornado is like the water in your bottle. Warm air rises, but when it's trapped below cold air, it can't rise easily. The warm air in a tornado is like the air in your tornado bottle. If the winds are curved just right, the warm air starts gushing up through a 'hole' in the cold air and a tornado forms. Wind speeds in the most destructive tornadoes can reach over 500kmh! The pressure in the eye of a tornado is also much less than the surrounding air. If the tornado passes over a house, the sudden drop in pressure outside can cause the house to literally explode.

On the news we see tornadoes in the United States every year, so some people are surprised to learn that they occur in Australia, too. Over 350 cases have been recorded in New South Wales alone! The one in this picture touched down around dusk on 18th March, 2007 near Harwood on the northern NSW Coast. I know because I took the photo myself. It only lasted about ten seconds and luckily didn't damage any houses. It's one of the most exciting things I have ever seen!

15 Hero's Fountain

Crafty little holes make water spurt up as it pours down this clever fountain. Challenge a friend to figure out why and you'll discover it's as cunning as it is fun to watch.

You'll need two straws, poster putty, strong tape, a drawing pin and two empty plastic bottles (1 or 2 litres will work best)

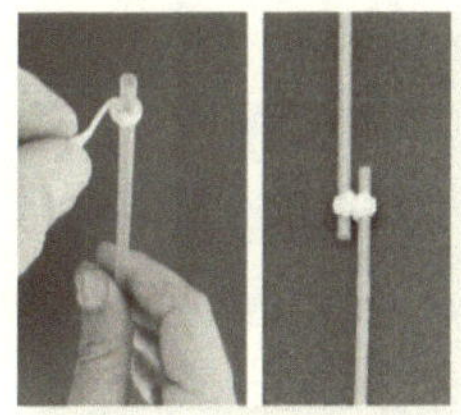

Stretch a small piece of poster putty. Wind it around a straw about 2cm from the end. Repeat with a second straw. Then press the straws together like this.

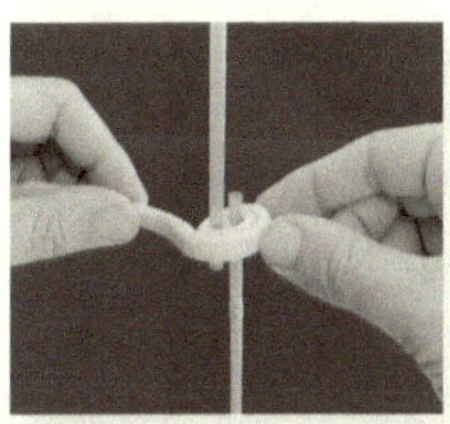

Now wind a larger blob of poster putty around the putty that's joining the two straws together. It needs to make a seal so it should be a bit wider than the mouth of your bottles.

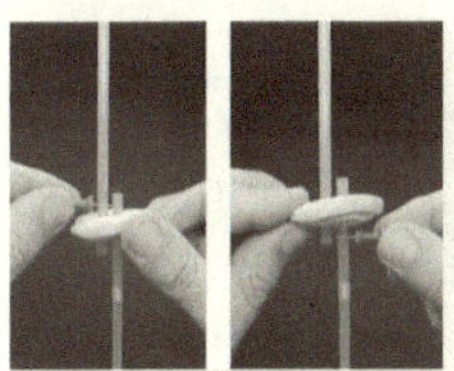

Carefully push a drawing pin right through the longer part of each straw. Each straw should now have two small holes on either side of the putty.

Fill one of your bottles with water so it is about three quarters full. Now press the straws onto this bottle.

Align and press the empty bottle down onto the first. Press the top one down firmly to make a good seal. Now use strong tape to hold the bottles together.

Your Hero's Fountain is ready to go. Gently turn it upside down and watch what happens!

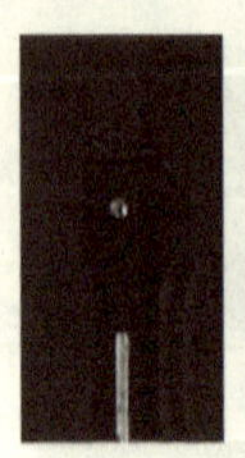

Watch carefully and you'll see water spurting up out of the top straw like this. Some drops will even hit the top of the bottle. Can you figure out why it happens?

What's going on?

When you turn the fountain upside down, the water starts flowing into the bottom bottle. You'll notice much more water pours through the straw that's poking into the bottom bottle. Let's call this the water straw. What you can't see is that an equal volume

air is pouring up into the top bottle through the other straw. Let's call this the air straw. A tiny bit of water trickles into the air straw through the tiny holes too. The upward air flow through this straw makes the water spurt out through the top.

Who's Hero?

Hero (or Heron) of Alexandria was a famous Greek mathematician and engineer. He lived from around 10AD to 70AD and was renowned as an excellent experimenter. He invented some truly amazing gadgets and contraptions.

Hero's engine was the first recorded machine that used steam. How it worked was not fully understood at the time. It was used to open temple doors and mystify worshippers. Hero also invented the first vending machine, and it dispensed 'holy water'. One of his most amusing inventions was a mechanical puppet show. The characters were moved by a system of ropes and pulleys operated by a large rotating cylinder.

When Hero wrote his *Pneumatica* he included the first recorded description of the water siphon, too. Hieroglyphics suggest that the ancient Egyptians knew how to siphon well before Hero was born, but they didn't record the way siphons worked. The lesson here is if you want to get the credit for a good idea, make sure you write it down!

16 Magic Bottle

Here's a classic old science trick that's been entertaining kids for years. For smaller hands, use a smaller bottle to make holding it easier. Please remember to conserve water while playing around with this trick.

Use a drawing pin to puncture a tiny hole near the top of a plastic bottle. It's easier if the bottle is fullof water (lid on). Cover this hole while you make another near the bottom of the bottle.

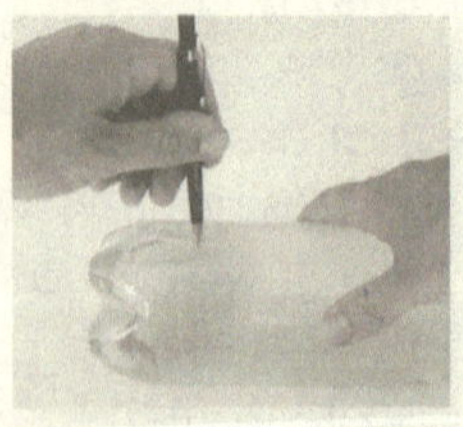

Push the tip of a pen or pencil into each hole to enlarge them a bit. Make them about half a centimetre wide. Now cover both holes with your fingers and turn the bottle upright.

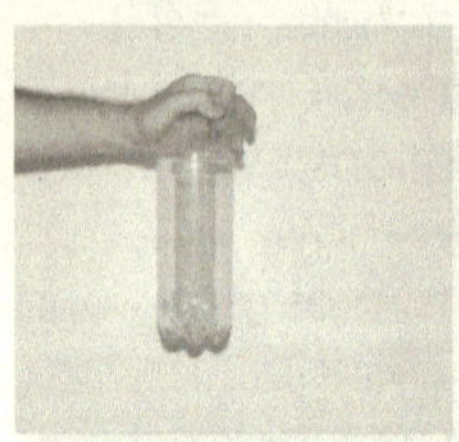

You only need to seal the top hole to keep the water inside the bottle. It won't leak – the water is stuck inside!

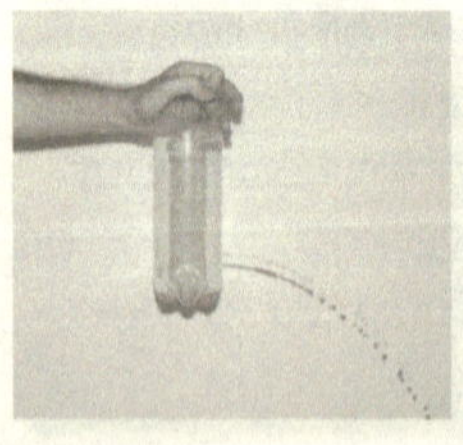

Take your finger off the top hole and water pours out. Cover it again and the leak stops. Do this secretly and people will think your bottle is magic!

What's going on?

Don't be too disappointed if most people already know the Magic Bottle's secret. This trick has been around for a very long time!

Air pressure and water's amazing ability to stick together combine to make your bottle 'magic'. For water to leak out of the bottle, an equal volume of air has to get in. With the lid on and your finger covering the top hole, the only way in is through the bottom hole. But water's amazing affinity for itself prevents air bubbles sneaking in.

The other reason no water pours out is air pressure. For water to pour out, the volume of air inside the bottle has to increase. Because no air can get in, this would cause a vacuum to form in the bottle. But the pressure outside the bottle is so great that it prevents this from happening.

Coffee Cup Lids

Inspect a takeaway coffee cup lid up close and you'll notice there's a big hole for drinking and a tiny hole on the other side.

The little hole is called a breather hole. It acts just like the secret hole in your Magic Bottle. When you drink through the larger hole, you also seal it. The breather hole lets air in, so the coffee can come out. Without it, no air could get into the cup to replace the coffee pouring out. The consumer would struggle to get a single drop out of the cup! There'd be bedlam in the streets. Without their morning cuppa, the world's coffee-drinking population would grind to a halt. The global economy would collapse as workers struggled against the invincible vacuums. I guess they could just remove the lid and drink straight from the cup. But that kind of ruins my dramatic ending! Oh well.

17 Inflation Impossible

There's still more magic in your Magic Bottle. Keep the holes in your bottle secret to convince people you have extraordinary lung power. They'll huff and they'll puff while you giggle and laugh.

Insert a balloon into your Magic Bottle. Stretch the neck of the balloon back over the thread of the bottle like this.

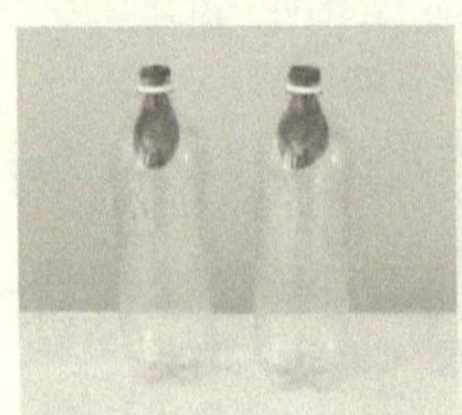

Do the same with a second bottle that has no holes in it. Give this bottle to a friend.

Ask your friend to inflate the balloon inside their bottle. Without any holes, this is impossible!

Now inflate the balloon in your magic bottle. No problems! Ask your friend to try again. They'll still fail. Conclusion? You have extraordinary lung power!

What's going on?

This challenge is an old classic among scientifically minded practical jokers. The reason you can only inflate the balloon in the bottle with holes is air pressure.

When you insert the balloon, the pressure inside the bottle is equal to the pressure outside. But the balloon seals the bottle so no more air can get in or out of the one without the holes. To inflate the balloon, you would need to compress the air between the balloon and the bottle. The air pressure inside the bottle is so great that you simply cannot inflate the balloon against it.

The reason it's so easy with your Magic Bottle is because air can escape though the holes. When you let the air out of the balloon and it shrinks, air rushes back in through these holes again. If you want to see the air escaping, wet a finger with soapy water and run it over the holes. With a soap film stretched over the hole, a little bubble will form when you inflate the balloon. If you seal the holes with sticky tape while the balloon is inflated, it stays inflated – and that looks pretty cool.

Vacuum Moulding

Did you notice how the balloon takes on the shape of the bottle when you inflate it? Another way to inflate it would be to 'suck' the air out through the small holes. Unfortunately, plastic drink bottles collapse if you try this because they're not rigid enough. If they were stronger, it would work. In fact, this is exactly how some types of plastic packaging are moulded. Rigid moulds are made with tiny holes in them. Sheets of warm plastic are laid on top. When the air gets 'sucked' out, the plastic takes on the shape of the mould. When it cools down, it stays in this shape.

18 Fibre Optic Water

Use your Magic Bottles and a torch to trap a beam of light! It's a bit messy, so do this one outside or somewhere safe. Catch the water in a bucket so you can use it again.

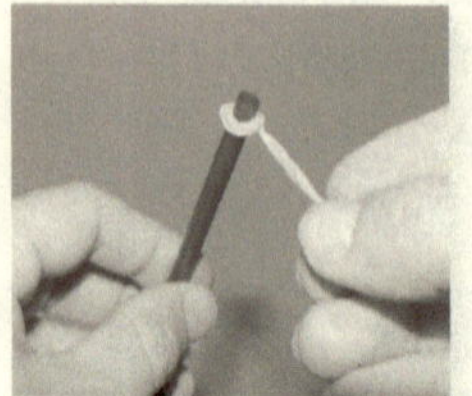

The stream of water needs to be nice and smooth for this trick. Adding a straw will help make it smooth. Stretch a small blob of poster putty. Then wind it around the end of a straw.

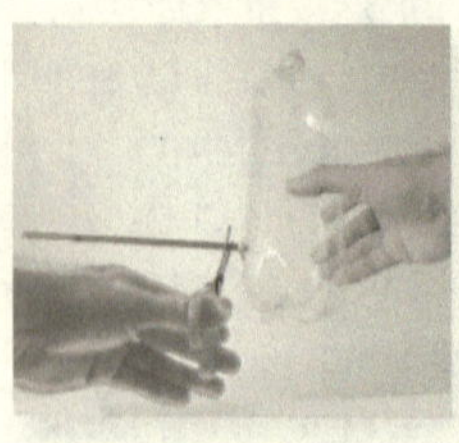

Insert the straw into the bottom hole and press the putty into the hole to seal it. Then trim the straw so it's only just poking out of the bottle like this picture.

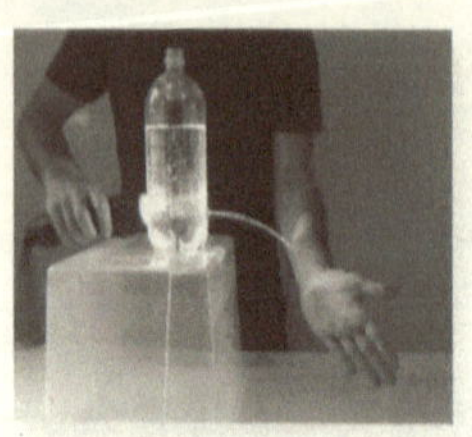

Let the water pour out onto your hand. Now shine a torch beam through the water in the bottle into the hole. Until you align the beam with the stream of water, not much happens.

When you align the beam just right, a bright spot appears on your hand. The light beam is following the bent stream… amazing!

BEACH
CLOSED

serious
experimentation
underway

What's going on?

To see why the stream traps light, try this fun experiment. Put on some goggles and slip into a swimming pool. You need a calm surface, so don't dive in or splash to much. Duck down about a metre, look directly up and you'll see the sky and clouds. The surface is acting like a window. Further away from you, you can't see the sky. Instead, you'll see a reflection of the bottom of the pool. Here, the surface is acting like a mirror. What you are seeing is called a total internal reflection.

Light beams can't emerge from water below a critical angle. It's because light travels a bit faster in air than in water. This is why your torch beam gets trapped in the stream. The beam keeps getting reflected back into the stream until it reaches your hand. Pretty illuminating huh?

Light up your MP3

If you love downloading songs and videos then you love optic fibre technology. Downloads stream into your computer as ones and zeros. Not the numerals 1 and 0 though – in an optic fibre, a one is represented by 'light on' and a zero by 'light off'. Real optical fibre is made from very thin strands of glass. By switching tiny lasers on and off really fast, music and videos are converted into signals made out of light and 'no light'. Those signals race around the world through optical fibres at the speed of light and phoomba – into your computer!

Optic fibres can carry heaps more of this 'on/off' information than electrical wires. Songs and videos would take heaps longer to download if we only had electric wires. So without total internal reflections, the internet would just be a big old binary snail.

19 DIY Magnifying Glass

A splash of water in the top of a soft drink bottle makes a surprisingly powerful little magnifier. You can use it to see enlarged images of anything you view through it.

Ask an adult to cut the top part off an empty soft drink bottle. The bottle needs to have smooth curved shoulders (not the conical kind).

Next, ask your helper to cut a section out of the top part like this, so it's like a scoop. It doesn't matter if it's not a very neat cut. Trim with scissors if you like.

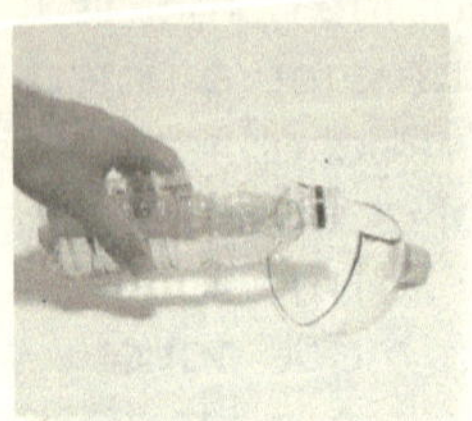

Pour a little bit of water into the scoop. The pool of water will act as a lens.

Hold your scoop of water over a page that has some writing on it. Whoa! You've got an amazing water magnifying glass!

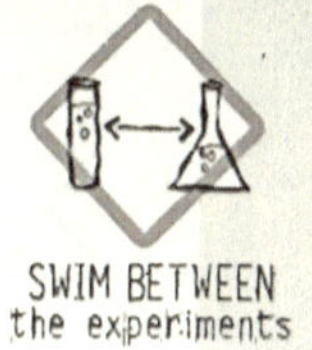

What's going on?

The top part of a soft drink bottle is curved like a lens, but it won't magnify images unless it has water in it. That's because a lens needs to have different curves on its front and rear surface. The plastic alone has a uniform thickness with the same curve on its front and rear surface. Add water and you are making a lens. Water is a liquid so the top surface becomes perfectly flat. The bottom surface however, takes on the curved shape of the bottle. The result is a superb little magnifying glass! Real magnifying glass lenses have two convex surfaces. Your water version has one convex and one flat surface.

Glassy facts

The magnifying glass was invented by a famous 13th century scientist and philosopher called Roger Bacon.

The main ingredient in glass is silicon dioxide (quartz). It is the most abundant substance in the Earth's crust. Pure silicon dioxide melts around 2300ºC. Adding sodium carbonate reduces the melting point down to around 1500ºC. Adding other chemicals can strengthen glass, change its colour or make it more heat resistant.

But the idea that the glass in windows slowly flows like a liquid over hundreds of years is not true. The story started when people noticed that old windows are often thicker at the bottom. It seemed like the solid glass was slowly flowing down. But early glass making was not as advanced at it is today, so window panes were rarely uniformly thick. For stability, glaziers always installed windows with the thickest part at the bottom. This simple explanation makes one thing crystal clear though: it's very easy to jump to the wrong conclusion!

20 Light Bender

You can use this tricky gizmo to decode your own 'secret' messages. You can use it to see how easily people jump to conclusions, too. Ask them why white letters on black backgrounds flip upside down while the black letters don't. Hmmm...

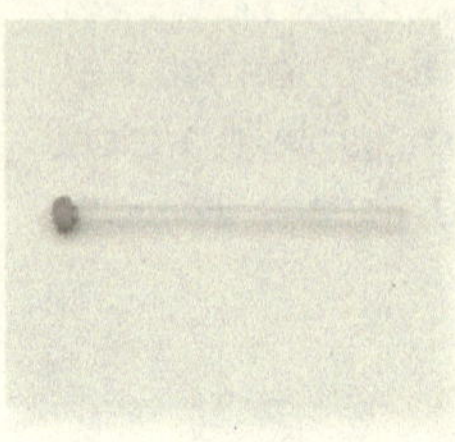

Seal one end of a clear straw with a blob of poster putty. Some fresh juice shops provide clear straws – ask politely and you might get one free (the word 'please' really helps!)

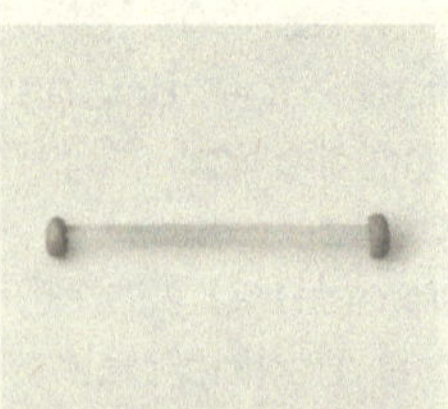

Fill the straw to the brim with water. Now seal the other end with a second blob of putty. Make sure there's no air left inside the straw.

Lay the straw over the secret code on the opposite page. Slowly lift the straw off the page while you look through it. A secret message will be revealed!

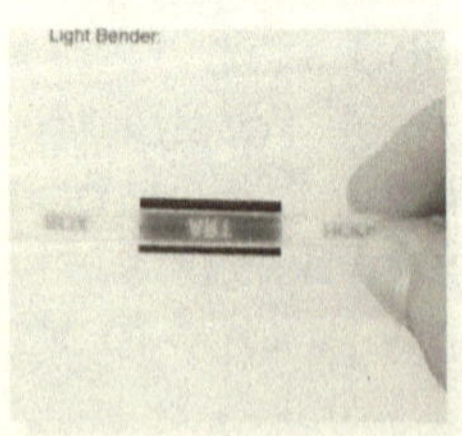

Now try it on the words below the code. Most people will guess the white letters get flipped upside down because of their 'colour'. Read on to find out what's really happening going on!

SCIENTISTS
patrol these waters

What's going on?

Your decoder is really just a cylindrical lens. Light rays passing through it get bent in a special way. If you hold it at just the right distance from the page, the image you see through it is flipped upside down. This is obvious with letters that do not have vertically symmetry like the ones below (try looking at them through your straw):

A a b F f G g h i J j L M m N T

But look at the capital letters below and you'll notice these don't appear to flip upside down:

B C D E H I K O X

The reason is that these letters all have vertical symmetry. The image is being inverted just like the others, but you don't notice because these letters look the same either way. It only works with very plain fonts though. Fancy fonts with swirls or serifs (the additional strokes at the end of the main stroke) like the ones below don't have vertical symmetry:

B C D E H I K O X

Now you know how it works, you can practise writing secret messages like the one below. You can also trick people with the white and black words below. Ask them why they think the white letters flip while the black ones don't. They'll probably say it's something to do with the colour. Sneaky, huh?

Use your decoder to look at these letters:

K66b FH12 CO96 26CK6F

BOX **ART** **HOOD** **SUN**

21 Wave in a Bottle

The Wave Bottle makes a cool and fascinating desk toy. Best to put it away at homework time, though. It's so much fun, you'll probably forget all about your spelling test and that would spell d-i-s-a-s-t-e-r!

Half fill a small empty bottle with water and add some food dye (blue looks cool). I found this fancy bottle at my local supermarket.

Next, pour in cheap vegetable oil until the bottle is full to the brim. Be very careful not to the spill oil – best to do it outside in the garden.

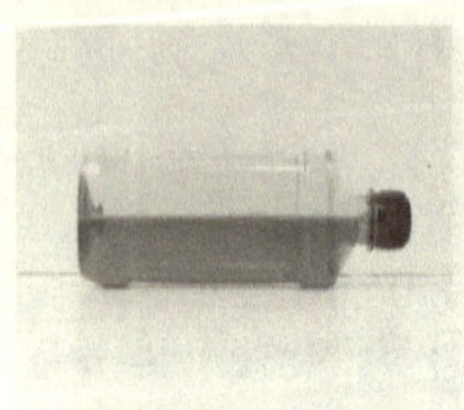

Dry any oil or water on the outside of the bottle thoroughly. You're ready to start making waves.

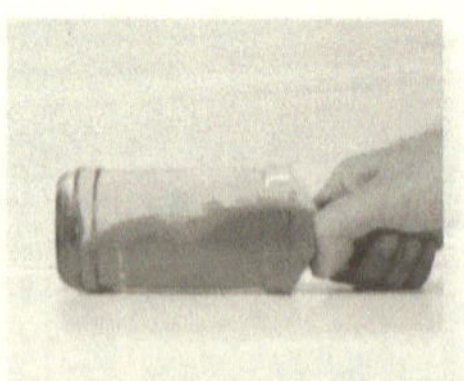

Tilt the bottle up and down and waves of water roll from one end to the other. Even if you break the oil into tiny droplets by shaking the bottle, it will eventually separate into two layers again. Cool!

Sandologist ON DUTY

What's going on?

Liquids that mix evenly together are said to be miscible. Miscible liquids form a homogenous (even) mixture in all proportions. In other words, one will always dissolve into the other. Water and oil however, will not mix. Scientifically speaking, water is immiscible in oil – and vice versa.

Oil is also a little less dense than water. A litre of pure water at 40° C weighs exactly one kilogram (1000 grams). A litre of oil weighs about 910 grams. That's why oil floats on water.

Even if you shake up your Wave Bottle, the oil and water eventually separate again. Vinegar and oil are immiscible, so they'll separate too. That's why you need to shake salad dressings before pouring them. The process of droplets joining up is called coalescence. But you can get oil to stay evenly mixed in water by adding a special type of chemical called an emulsifier. Such mixtures are called emulsions. Emulsifiers are chemicals that can mix with both oil and water. Egg yolks contain natural emulsifiers, which are whisked together with oil to make yummy mayonnaise. It's super delicious on hot chips, but go easy – mayonnaise is also super-fattening!

White Water

The waves in your bottle don't make white foam like the waves you see at the beach. That's because there's no air inside the bottle. The white water that ocean waves make is white because it contains billions of tiny air bubbles. Each bubble reflects a little bit of light. Together, they reflect so much light in every direction that the water looks white, even though it's actually clear!

22 Micro Surfer

Here's a fun little addition for your Wave Bottle. Make the Micro Surfer so it floats in water but sinks in oil. Then it will surf the water layer in your Wave Bottle.

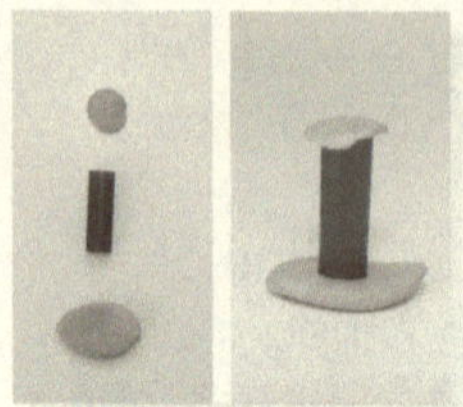

Cut a straw to make a short piece about 1.5cm long. Use a tiny blob of poster putty to seal the top and a larger blob for the bottom. Shape the bottom blob like a tiny surfboard.

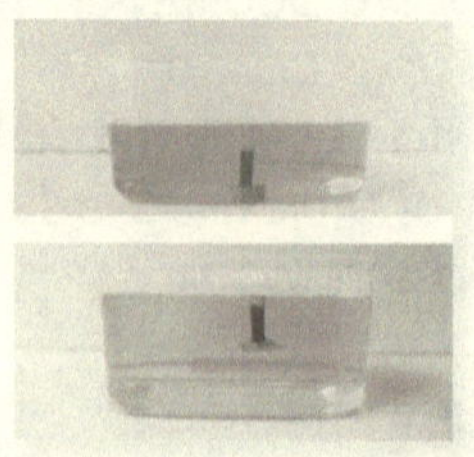

You need your Micro Surfer to only just float in water. Make sure yours sinks like the one in the bottom picture first. Then gradually remove tiny bits of putty until it just floats like the one in the top photo.

Put the Micro Surfer carefully into your Wave Bottle. Seal and dry the bottle thoroughly. Now make some waves.

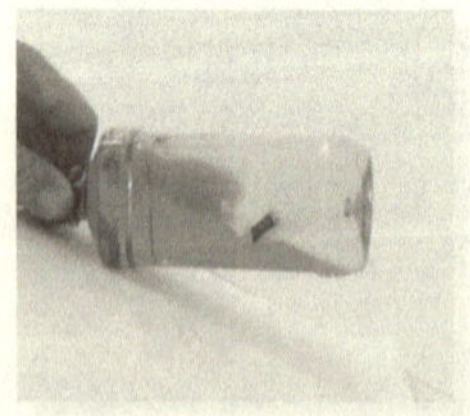

With a bit of imagination, your Micro Surfer really rips. This one is about to get tubed!

BEACH CLOSED

serious experimentation underway

What's going on?

Any object is buoyant in water if the water it displaces weighs more than the object. Sound confusing?

Imagine pushing a tennis ball deep into a jug that's full to the brim with water. The water that spills over would weigh much more than the tennis ball. A tennis ball floats because it weighs less than the water it displaces. Cricket balls are heavier, but they too weigh less than the water they displace. So, cricket balls can float too. But golf balls weigh more than the volume of water they displace. So golf balls sink!

Now here's the trick. An equal volume of oil weighs a bit less than water. As a result, objects will be slightly less buoyant in oil than in water. The Micro Surfer needs to float in water, but sink in oil. That's why you should start with a Micro Surfer that sinks in water and only remove small bits of putty until it only just floats.

Eureka!

The link between weight, volume and buoyancy is called Archimedes' Principle. He was a Greek mathematician and scientist born around 287 BC. Archimedes was asked to determine if his king's new crown was a fake. He knew that pure gold weighs more than fake gold. But without melting it down, there was no way of knowing the crown's exact volume for comparison. One day, he noticed the water level going up as he got into the bath and his great idea struck. He would submerge the crown in a jug full of water. The volume of water that spilled over would be exactly equal to the crown's volume. Legend has it he jumped out of the bath and ran through the streets yelling 'Eureka! Eureka!'. It means 'I've found it!'. And yes, the crown was indeed a fake!

23 DIY Lava Lamp

The secret chemicals inside Lava Lamps have earned millions of dollars. They're famous too. There's one on TV, somewhere, nearly every night! Now if that's not enough to get you excited about chemistry, I give up.

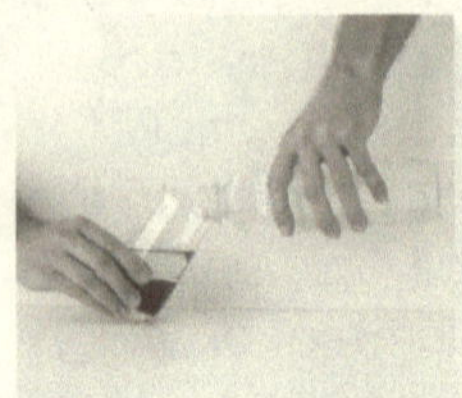

Quarter fill a glass jar with water. Add a few drops of food colour. Hold the glass at an angle and gently pour in vegetable oil until the glass is almost full.

Now gently drop in a soluble aspirin, vitamin or antacid tablet. Make sure you ask for permission first, of course!

Notice that nothing happens while the tablet is in the oil. Once it hits the water, though, it starts fizzing and the show begins!

Big blobs of coloured water float to the surface, then sink again. It looks just like a real lava lamp!

What's going on?

Tablets that fizz in water contain an acid and a second type of chemical called a base. While the tablets are dry, the acid and base can't react. Dissolve one in water however, and the acid-base reaction takes off, producing bubbles of carbon dioxide gas.

Water sinks in oil because it is denser. Carbon dioxide gas is less dense than oil so it floats. The rising blobs of water contain lots of carbon dioxide bubbles, so they float. At the surface, the bubbles pop, the gas escapes and the blob sinks again.

Use a jar with a lid so you can keep your lava lamp safe. Then you can use it over and over again. Use a teaspoon to carefully remove any floaty bits that build up on the surface.

Real Lava Lamps

The secret ingredients in a real Lava Lamp were invented by Edward Craven-Walker in the 1960s. The light bulb inside heats up the water and the dense goo. When it melts, the goo expands. Now the goo is less dense than the water, so it floats. When it cools down at the top of the lamp it shrinks, so it becomes denser than the water. Now it sinks again and the whole process repeats.

Edward got his Lava Lamp inspiration from another kooky invention – a weird egg timer! It contained a blob of solid wax and a liquid sealed in a glass tube. The whole thing went in the water with the eggs. When the molten wax began to float to the top of the trapped liquid, your eggs were cooked.

Edward spent fifteen years perfecting his chemicals before launching his Astro Lamps (the name changed later). Many millions have been sold since. Not bad for a kooky little lamp full of goo!

24 CD Hovercraft

This balloon-powered hovercraft glides on a thin cushion of air. Use a marker to decorate your balloon with cool racing stripes or flames for an even snazzier hovercraft.

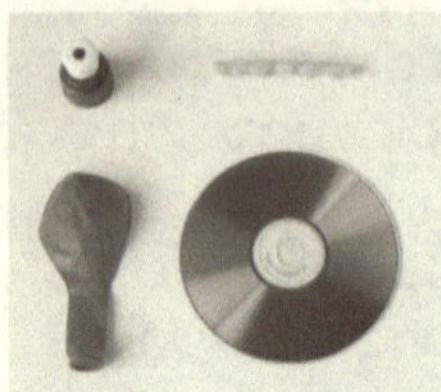

You'll need an old CD, a sports drink bottle nozzle, a balloon, and a big blob of adhesive poster putty.

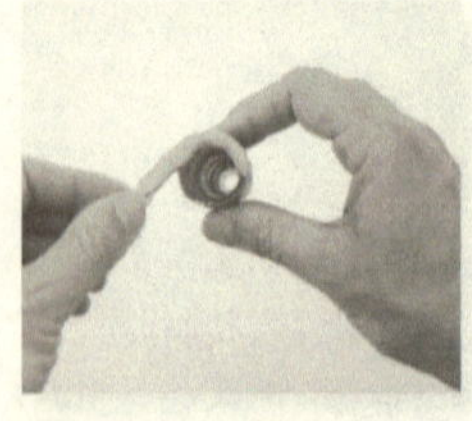

Make a fat worm out of adhesive putty. Wind it around the bottom of the nozzle.

Centre the bottle lid on the CD and press firmly to make sure it's airtight. To test for leaks, cover the hole in the CD and blow into the bottle top. No air should come out.

Stretch a balloon over the nozzle. Inflate by blowing through the hole in the CD. Pinch the balloon to keep the air inside. Put it on a flat surface and give a gentle push. Coolio!

What's going on?

Friction usually brings a sliding CD to a stop. Friction is the force between two surfaces that makes sliding difficult. The air rushing out of the balloon lifts the CD slightly so it's not actually touching the table. Now the CD is sliding on a layer of air. The friction between the CD and air is much, much less than the friction between the CD and the table. With less friction, the CD glides gracefully across any smooth surface.

Real Hovercraft

The hovercraft was invented in England by Sir Christopher Cockerell. He was an engineer and a tireless inventor. He came up with the idea in the 1950s, while thinking about how to make boats go faster. He realised that friction was a major limiting factor.

He began experimenting with empty cat food and coffee tins, an industrial air blower and kitchen scales. His hover theory worked and he soon had a working model levitating on a thin cushion of air. He took this prototype to Whitehall in London. The expert onlookers were highly impressed, but wanted to keep the invention secret for military use. In 1959, Sir Christopher convinced the government to let him build a commercial hovercraft for civilian use. Within weeks of its launch, the hovercraft made its first successful English crossing.

Since then, hundreds of millions of people and thousands of cars have travelled by hovercraft. Sir Christopher was knighted in 1969 for his services to engineering. And just think – it all started with a great idea and some empty cans in a humble kitchen!

25 Pop-A-Rocket

Pop-A-Rockets are powered by a puff of air and can soar over five metres high. Use the design on Page 93 to add paper fins and you've got yourself a miniature carpet-to-air missile program.

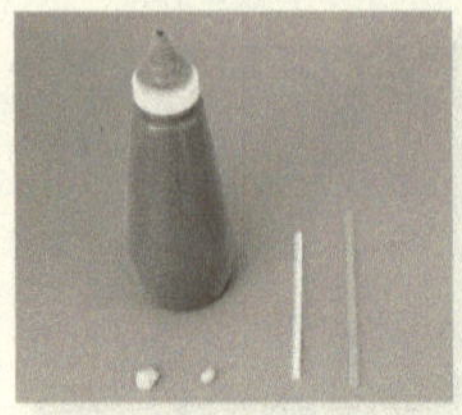

Keep an empty squeeze-type sauce bottle. You also need adhesive poster putty, a thick straw and a slightly thinner one. Cut the thicker straw so it's a bit shorter than the thin one.

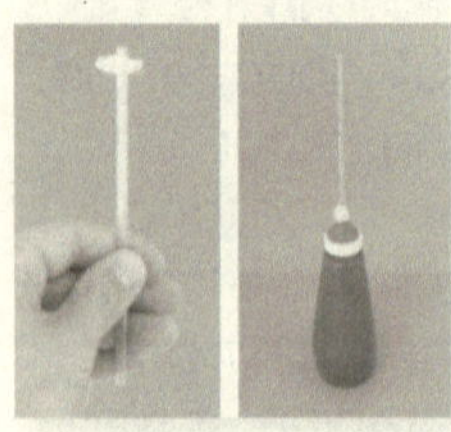

Wind some poster putty around the end of the thinner straw. Screw the bottle nozzle so it's open and push the straw into the hole. Press the putty down to seal the straw onto the bottle.

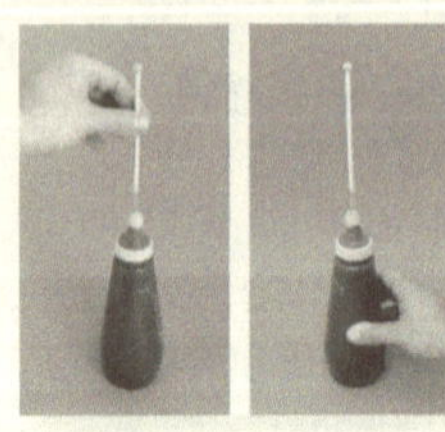

Seal one end of the thicker straw with a small blob of poster putty. Slide it over the thinner straw on the bottle. Squeeze hard and the rocket goes flying.

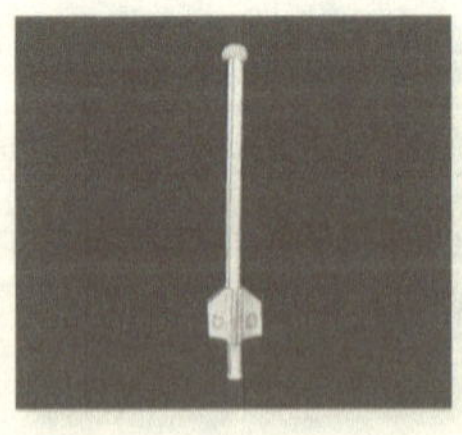

To make fins, photocopy the design on page 93. To attach your fins to the straw, follow the instructions and you've got yourself a cool little rocket.

Sandologist
ON DUTY

What's going on?

The Pop-A-Rocket demonstrates a phenomenon called Newton's Third Law. This famous law states that 'for every action, there is an equal and opposite reaction'.

When you squeeze the bottle, some of the air inside gets forced out through the straw. Whenever a large volume of air gets forced through a narrow tube, it rushes through at great speeds. This air collides with the sealed end of the thicker straw and bounces back down again. The collision between the air and the sealed end of the larger straw generates the force that makes it fly. The 'action' is air bouncing off the poster putty. The 'reaction' is the Pop-A-Rocket shooting into sky at great speed.

Real rockets

The earliest rockets were powered by gunpowder, which was invented in China. They were used to launch fireworks – and they're cool. They had devastating military applications, too – and that's not cool!

The first report of human rocket-powered flight dates back to 1633 in Ottoman Turkey. According to the history books, Lagari Hasan Çelebi launched himself some three hundred metres into the air. He used gunpowder as the rocket fuel and his flight lasted about just over three minutes.

Rockets have come a long way since then and can go a lot further too. Unfortunately, most rockets are used to carry deadly explosives. They're amazing to watch, but you don't want to be standing nearby when they go off! Missiles are ridiculously expensive, too. A single Tomahawk missile costs about one million dollars! You could buy more than two thousand brand new surfboards for the same money – cool? Nah, crazy!

26 DIY Juggling Balls

Rice-filled balloons make excellent juggling balls. The problem is getting enough rice into the balloon. Here's a clever, fun and simple way to do just that. Now all you have to do is learn how to juggle!

Pour a cup of uncooked, dry rice into an empty bottle. Inflate and stretch a balloon over the bottle. Turn the bottle upside down and all the rice falls neatly into the balloon.

Remove the balloon carefully and let it slowly deflate. Use scissors to trim the neck off the balloon. Be careful not to spill the rice.

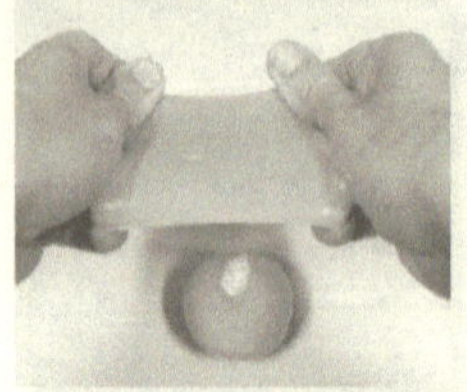

Cut the neck off a second balloon. Stretch it over the hole in the first. The rice is now safely sealed inside. Now it's time to add some decoration.

Cut the neck off a third balloon and add some extra holes for decoration. Stretch it over the ball. Repeat to make two more balls and hey presto – you're ready to juggle!

BEACH CLOSED
serious experimentation underway

What's going on?

Rice is an excellent filler for juggling balls because it's nice and heavy. Small, heavy balls make learning to juggle much easier. Now you've got some excellent juggling balls, the only thing left is to learn how to juggle. It's actually quite easy if you follow the simple steps below. It only takes about half an hour to learn so get cracking!

Juggling tutorial

BIG HINT: Practise over the end of a bed. If you drop one of the balls, you won't have to reach down so far to pick it up! This saves lots of time, so you can get straight back to your juggling.

STEP 1: Start with two balls in your writing hand, and one in the other. Toss one of the two balls up and across to your other hand. Catch it and stop there. Pass (don't throw) this ball back to your writing hand again. Practise this until you can do it without looking. It might take five to ten minutes of practice. Don't rush!

STEP 2: Start with two balls in your writing hand and one in the other. Toss one of the two balls across to your other hand like before. While it's still in the air, toss the ball in your other hand up and across to your writing hand. Stop there! Practise this for five or ten minutes. Don't rush this step!

STEP 3: Now you're ready to try juggling. Repeat step two, but this time throw the third ball in your writing hand before catching the one in the air. If you're struggling, practise Step 2 again until you can do it with your eyes closed.

Once you've mastered juggling balls, you can move straight onto chainsaws – JUST KIDDING!!!

27 Cartesian Diver

This gizmo is named after the French scientist and philosopher René Descartes. He's also famous for saying, 'I think, therefore I am'. This diver makes me suspicious. Maybe he really said, 'I sink, therefore I… swam'?

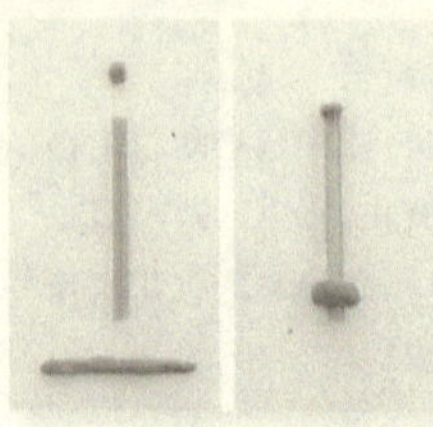

Cut a 6cm length of drinking straw. Use a small ball of adhesive poster putty to seal one end of the straw. Use a bigger blob to make a long worm of putty. Wind it around the other end of the straw.

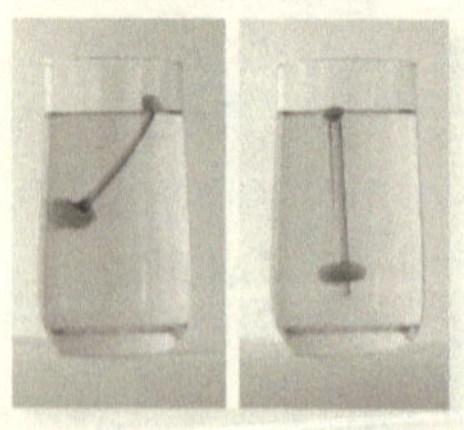

Test your diver's buoyancy in a glass. If it floats on an angle, it's way too buoyant and needs more putty. If it sinks, remove some putty.

Pop your straw diver into a bottle full of water. Make sure the bottle is full to the brim. Screw the lid on tight.

Now squeeze the bottle and the diver sinks. Stop squeezing and it floats back to the top. Look carefully and you'll figure out why.

What's going on?

The straw floats because the air bubble inside provides buoyancy. When you squeeze, it sinks because you are squashing that little air bubble inside.

Try squashing a bottle full of air – easy! Now try squashing a bottle full of water – impossible! Scientifically speaking, air is a compressible fluid. Water on the other hand, is incompressible.

Squeezing the bottle reduces the volume inside. If you squeeze it with the lid off, water pours out. With the lid on, the only thing you can squash in the bottle is the air inside the straw. When you squeeze, water rushes into the straw and the air bubble gets compressed. This reduces the straw's buoyancy, so the straw sinks. Stop squeezing and the water rushes out, the bubble expands and the straw rises again.

If you squeeze with just the right pressure, you can get the straw to hover anywhere you like. At this pressure, the straw is neutrally buoyant. It neither floats nor sinks.

Submarines

Air is used to change the buoyancy of real submarines, too. Large chambers can be filled with water or air to sink and surface.

The deepest dive ever accomplished by human beings was aboard a submarine called the Trieste. On 23rd January 1960, Jacques Piccard and Lieutenant Don Walsh reached the Challenger Deep. At almost 11km deep, this is the deepest surveyed point anywhere on Earth. No light penetrates this murky world. The pressure is so great down there it would crush a conventional submarine flat. Yet even at this incredible depth, Piccard and Walsh saw several fish swim by! Nobody has returned to this amazing place since.

28 Obedient Octopus

This little critter really looks like a tiny octopus. You can make cool stick-on eyes out of a white bread clip. Use a hole-punch to cut them out and a permanent marker to add pupils. Ta-da! Googly octopus eyes!

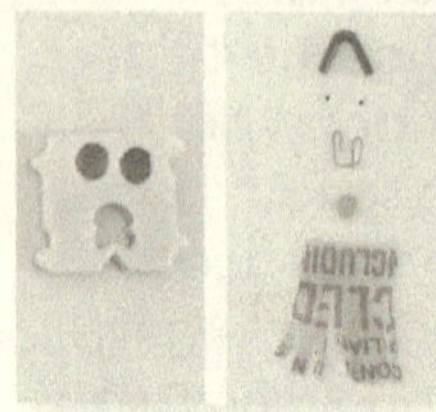

Cut the flexible part from a bendy straw. Bend a paper clip into a U shape. Cut slits in a rectangular piece of plastic bag for tentacles. Pinch off a small blob of poster putty.

Thread the tentacles onto the paper clip. Push the ends of the paper clip into the ends of the straw. Press the putty onto the paper clip. Use putty to stick on the googly eyes.

Make sure your octopus only just floats. If its head sticks out too far, add some more putty. If it sinks, remove some putty.

Put your octopus inside a bottle full of water. Screw the lid on tight. Squeeze and your octopus sinks. Stop squeezing and it obediently rises. Good little octopus!

SCIENTISTS patrol these waters

What's going on?

The Obedient Octopus is really just a fancier Cartesian Diver like the one on page 62. It works the same way, but looks remarkably like a tiny octopus.

Octopus facts

TThe plural of octopus is octopuses, not 'octopi'. They probably don't care about grammar, but they do have the largest brains of all invertebrates (animals without backbones). Instead of fins, they use jet propulsion to move through the water. And if eight arms with suckers aren't weird enough for you, octopuses also have three separate hearts and blue blood. Their blood is rich in copper and that makes it blue. Mammal blood is rich in iron, which makes it red.

Those big brains make them fast learners and excellent problem solvers. Intelligence is a hotly debated topic though, so scientists have put octopuses through all kinds of tests. These usually involve a food trap or maze. By changing a maze or trap an octopus has already solved, scientists hope to understand how they 'think'. In 2003, an octopus named Frida made world news by learning to unscrew the lids off glass jars full of shrimp. They are also masters of disguise and great escape artists too. Some species can make the clouds of ink they squirt out look like themselves! This bamboozles predators while the octopus makes a snappy escape.

The giant Pacific octopus can grow to over five metres. The tiny pygmy octopus only grows to 2cm. They might seem alien and even frightening to us, but they're extremely shy too. So remember, with red blood, and two sucker-less arms and legs, you're probably as weird to an octopus as it is to you – you backboned weirdo!

29 Straw Siphon

This gizmo's for suckers. No, really – because you need to suck the straw to get it going. It's more fun if you add some food colouring and use translucent straws so you can see the water clearly.

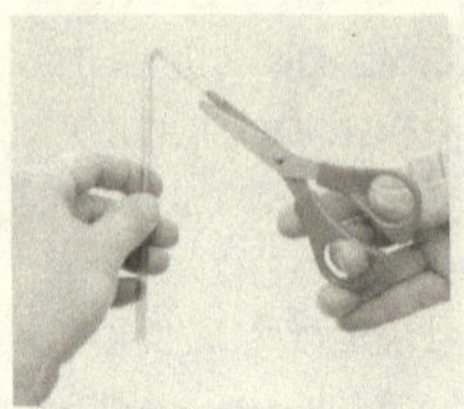

Cut a small slit in one end of a bendy straw. Push this end of the straw into a second bendy straw.

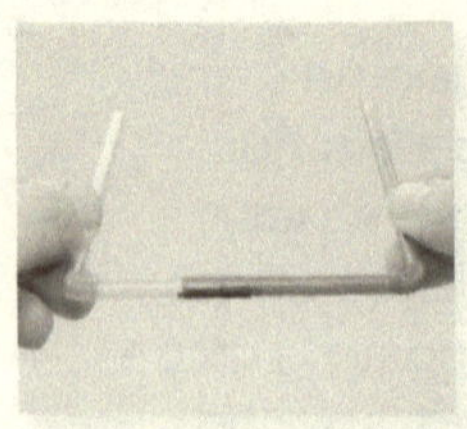

Seal the join with sticky tape. Keep joining straws until you have a super long bendy straw.

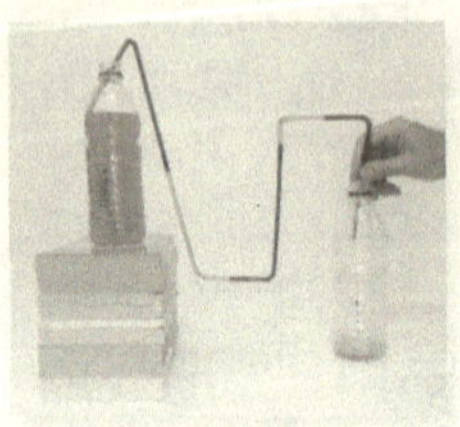

Fill a small bottle with water. Put one bottle on a stack of books or something similar. The end of the straw in the empty bottle should be lower than the one in the full bottle.

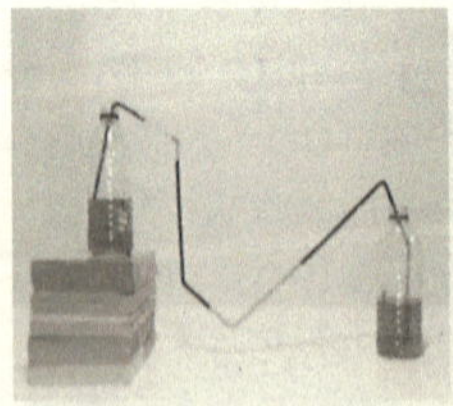

Suck the straw to get the siphon going. This is called 'priming' the siphon. When water comes out, put this end in the empty bottle. The water will keep flowing until the higher bottle is empty.

Sandologist ON DUTY

What's going on?

Siphoning is a very handy way to transfer liquids from one container, called the reservoir, to another. Once it starts, the liquid continues to flow, even if the middle of the hose is lifted high above the reservoir. Siphons work because the molecules in a liquid are strongly attracted to each other. All matter is made of molecules, but they're way too small to see – even with a microscope. They're just unimaginably tiny.

The individual grains in a dry powder like salt or sand can 'flow' too but they cannot be siphoned. There are two reasons for this. First, the individual grains are not attracted to each other like the molecules in a liquid. Second, the grains are too large and have tiny air spaces between them. Air bubbles in the hose will stop a siphon from working.

For a siphon to work, the end of the hose must be lower than the one in the reservoir. That means the hose must be longer on the lower side. So when the hose is full, the lower side contains more liquid than the reservoir side. Just as a scale balance always hangs lower on the heavier side, the liquid in the lower side of a siphon pulls down harder than the liquid in the reservoir side. This imbalance pulls liquid up and out of the reservoir, over the high point and down on the lower side. The molecules are so strongly attracted to each other that they won't let go and the siphon flows.

Siphoning Limits

There's a limit to how high the hose can be above the reservoir beyond which the siphon won't work. For water at atmospheric pressure, it's about 10 metres. That's almost four storeys high! Any higher and dissolved gases in the water will turn into bubbles and the siphon stops.

30 Mobile Phone Disco

This miniature disco really rocks. A mobile phone provides the music for the cling wrap dancefloor. Pack it out with salt and pepper or the tiny paper dancers on Page 93. Then pump up the tunes and watch 'em go!

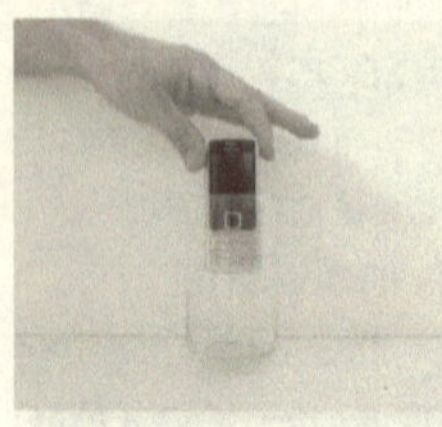

Select a cool polyphonic ringtone with a thumpy bass line. Turn off the vibrating alert and voicemail feature. Then put the phone inside a glass.

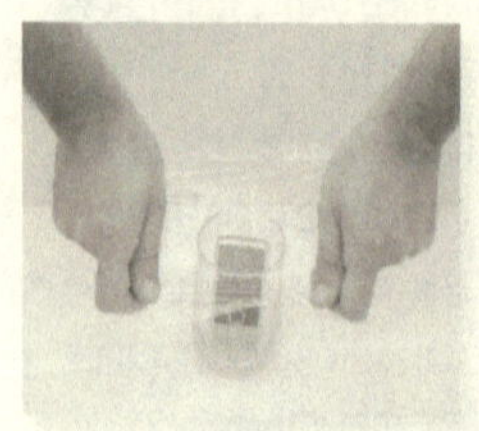

Stretch some cling wrap over the glass. Make sure it's nice and taut like the skin of a drum.

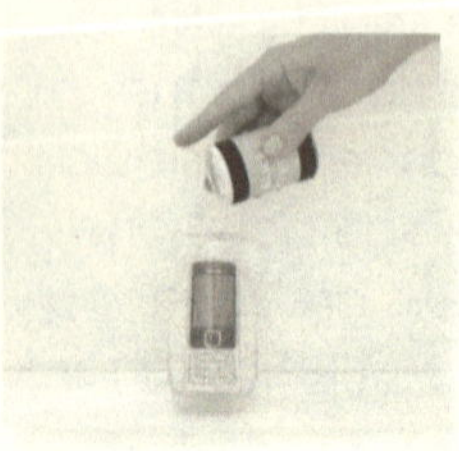

Sprinkle salt and/or ground black pepper on the cling wrap. For extra laughs, photocopy Page 93. Cut and fold the tiny dancers as instructed and set them up on the floor.

Use a second phone to call the mobile phone. When it rings, the salt, pepper and tiny dancers start bopping. Whoa – this place is pumping!

What's going on?

Sound waves are vibrations. You can't see them travelling through air, because air is invisible. When sound waves hit your cling wrap 'dance floor', it starts vibrating, too. The vibrations make the salt, pepper and paper dancers jump up and down. They literally dance to the tunes.

Lower frequency sounds, like the bass line of a good rock song, have slower vibrations than high-pitched sounds. The stretched plastic film has a natural vibration frequency that is fairly low. That's why the film responds best to bass lines. Bass lines usually set or follow the rhythm. That's why the salt, pepper and paper dancers bop up and down in time with the beat – very groovy, baby!

Eardrums

The human eardrum works just like your cling wrap dance floor. The eardrum is a membrane that is stretched across the ear canal. Sound waves travel into the ear canal and cause the eardrum to vibrate. A tiny little bone attached to the middle of the eardrum transmits these vibrations into the cochlea (the inner ear).

The cochlea is filled with liquid. The eardrum's vibrations travel into this liquid, causing tiny hairs inside to vibrate, too. An amazing thing happens in the base of these hairs. The vibrations get converted into electrical signals. These signals travel along nerves all the way to your brain. Your brain interprets the signals and you hear them as sound. Amazing!

But here's something everyone with an MP3 player should know. You should be able hear people speaking while listening to your music. Otherwise, you're probably damaging your amazing sense of hearing!

31 Slinky Spring SFX

This gizmo makes a super cool sci-fi laser cannon sound effect. You can use it like a string telephone too, but this one has reverb! So this gizmo sounds as cool as it – sounds!

You need two plastic cups, sticky tape and a small slinky spring. You can find these in the toy section at most bargain shops.

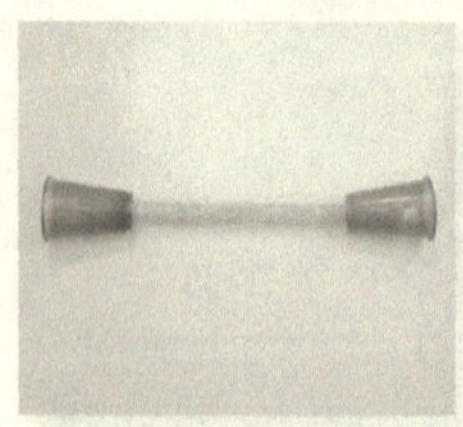

Use sticky tape to secure the ends of the slinky firmly to the bottom of each cup.

Grab a friend and put the cups to your ears. Now tap the slinky with a pencil or a fork. *Kerpow-ow-ow!* Stretching the slinky makes different sounds.

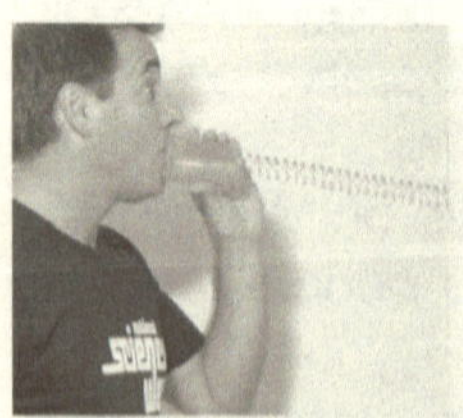

Speak loudly into the cup and your friend will hear a muffled, echoing voice. If your mobile phone has a voice recorder, put it inside the cup to record the sounds!

SWIM BETWEEN
the experiments

What's going on?

As you discovered with the Mobile Phone Disco, sounds are vibrations. Every single sound you hear is produced by vibrations. A cricket ball hitting a bat makes the bat vibrate. The vibratiing bat pushes the air molecules around it. Those air molecules bump into *their* neighbours. The neighbours bump into their neighbours – and so on. When the vibrations reach your ear, you hear the crack of ball hitting the bat.

Slinky springs always make the cool sounds you've just heard, but you don't normally hear them. That's because the coils are very thin. They have a very small surface area, so they only push against a tiny bit of air. The vibrations dissipate rapidly in air as they radiate away from the coils. The bottom of the cup has a much greater surface area, so it pushes a lot more air. This makes the vibrations much louder, so you can hear them. By sticky taping the slinky to the cup, you amplify the sound.

Speaking into one cup causes the bottom to vibrate, too. The vibrations bounce off the end of the slinky and race up and down many times. That's why you hear an echo. It sounds like someone speaking in a huge empty room.

Spring Reverberators

Show this trick to a guitar player and you'll see a light bulb flickering inside their head (not really, but you know what I mean!) Early guitar amplifiers used springs to create the reverb effect, too. Reverb is the cool echo sound you get in a big empty room. The sound reverberates off the walls, ceiling and floors. The reverb spring was usually stretched across the back of the amplifier. It would keep vibrating after the guitar strings stopped making sound.

32 Annoying Straw Trombone

Welcome to the most annoying gizmo in this book. The horrible sound it makes drives people nuts. It's worse than singing out of tune so no one can think straight. And that, of course, is exactly why you should make one right away!

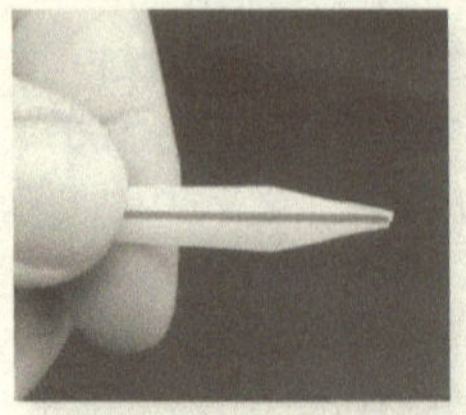

Flatten the end of a straw between your teeth. Use scissors to cut the end so it looks like a V. Snip the sharp tips off to protect your tongue!

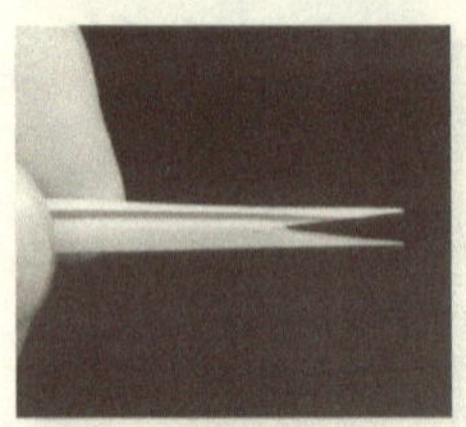

Your straw should look like this. To play, clench it tightly between your lips about 2 cm from the end. Blow and adjust the pressure between your lips until you get a loud sound. Eek – it's terrible!

Once you've mastered getting a sound, try cutting the straw while you blow. Each cut shortens the straw and the pitch goes up. It's even more annoying than before!

Slide a second, slightly fatter, straw over the first. Sliding up and down changes the pitch. You can even play a tune! (Pulling a ridiculous face like the one in this picture is optional, though.)

SCIENTISTS
patrol these waters

What's going on?

All sound is made by vibrating objects. Look at a guitar being played and you can see the strings vibrating. The tighter the string, the faster it vibrates and the higher the note. Shortening the string makes it vibrate faster, too. Playing a guitar is all about making the string longer and shorter by pressing on the fret board.

The Annoying Straw Trombone makes a sound because blowing through it causes the ends to vibrate. This is very similar to the way the reed in an oboe works. It's so similar in fact, that this gizmo should really be called the Annoying Straw Oboe!

When you blow, the vibrations travel down the straw and emerge as sound. The length of the straw affects the pitch of this sound. By cutting the straw, you shorten the vibrations, so the pitch goes up. You can't stick those bits of straw back on again, but sliding a second wider straw means you can shorten and lengthen the straw like a trombone. With a bit of practice, I managed to play 'When the Saints Go Marching In' on mine. It sounded horrible. I also managed to annoy my neighbours, so my advice is not to try this first thing on a Saturday morning!

Reed Instruments

The reeds in a reed instrument are made from cane and shaped especially for each instrument's mouthpiece. Clarinets and saxophones use a single reed. Oboes, English horns, bassoons, contrabassoons and bagpipes all use double reeds.

Reed instruments have a very distinctive sound and feature in all sorts of music, like film scores and television commercials. Even your favourite cartoon or computer game theme tunes probably have them!

33 Frozen Hand

This icy appendage looks awesome in bowl of punch. Everyone will want to meet the master sculptor who created it! You'll need to cut the glove, so please ask for permission before raiding the cupboard under the sink.

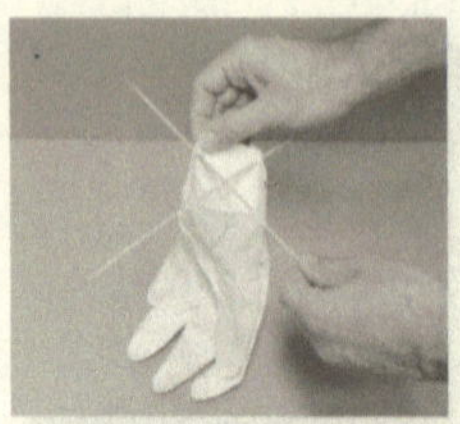

Poke two wooden skewers through the wrist of a rubber glove. The skewers should make a cross.

Suspend the glove inside a tall container. Fill with coloured water. Red is great, because it looks like freaky blood!

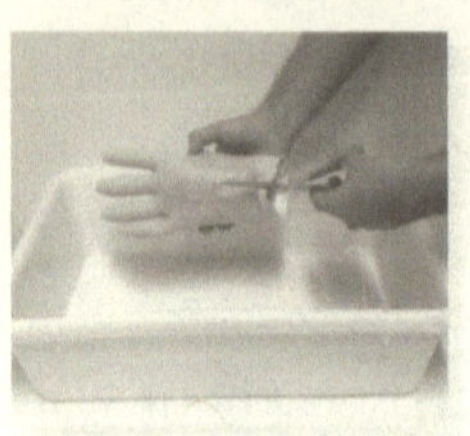

Put the glove in the freezer and leave it overnight. Remove and carefully cut the glove so you can peel it off. If you snap a finger, firmly press it back on. It should refreeze onto the hand.

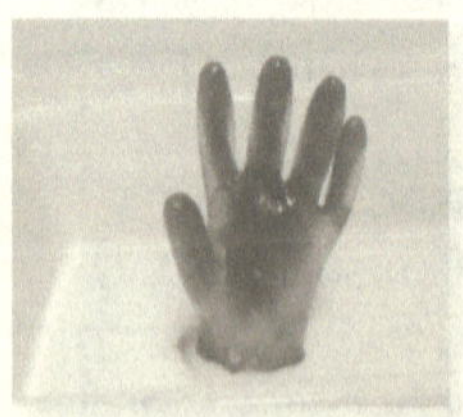

Your Frozen Hand is ready to shock and entertain. I shook hands with this one, but it gave me the cold shoulder – oh, *snap!*

What's going on?

Inspect your Ice Hand closely and you'll notice some very interesting things. The food colouring you poured in was nice and evenly mixed. Upon freezing, however, the surface ice becomes clearer. The food colouring gets concentrated in the middle. There's a good reason for this.

Pure water freezes at 0°C but tap water is not pure. It has lots of minerals and gases dissolved in it. These and the food dye are called impurities. Water is a liquid above 0°C. Water molecules are strongly attracted to each other, but above 0°C they can't form lasting bonds. Heat energy jostles them about, so any bonds that do form immediately get broken again. It's like a very crowded ballroom where the dancers change partners every time they bump into someone. In liquid water, the molecules are changing partners billions of times per second!

When you cool water down, the molecules lose some energy and their 'dance' slows down. At 0°C, water molecules can stick to their partners permanently. This is called freezing. But water molecules are more attracted to each other than to impurities. They need to be colder than 0°C to form bonds with other molecules. It's as though water won't dance with food colouring unless the music is even slower.

In your freezer, the glove cools from the outside in. The dance slows down on the surface first. The impurities won't find partners here so they 'dance' further into the glove. But it's about –10°C in your freezer, so eventually the impurities get locked in the middle. The result is clearer ice at the surface with the impurities concentrated in the middle.

34 Haunted Glove

Here's a creepy way to blow up a glove. Cover your jar so no one can see what's inside and your glove looks like it's being haunted by the hand of a ghost. Ooooh!

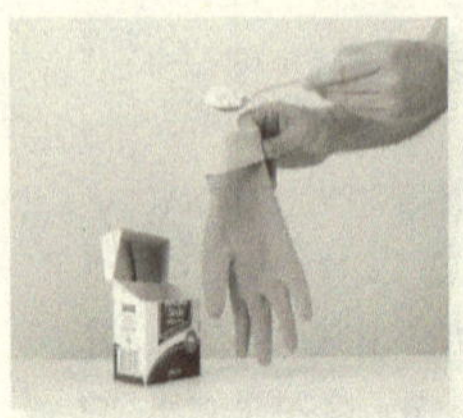

Getting these quantities right takes some experimenting. Start with a tablespoon of baking soda (sodium bicarbonate). Shake the glove so the baking soda falls right into the fingers.

Pour about 50ml of vinegar and 50ml of water into a large jar. Then stretch the glove over the mouth of the jar. You might need to secure it with a rubber band.

I've used a glass jar so you can see what's going on inside. Wrap paper around the jar if you want to keep it secret. Draw a skull and crossbones to make it extra freaky.

Lift one of the fingers momentarily so some baking soda falls into the jar. Let go and the glove will slowly inflate like this. Eek – there's a ghost in that glove!

BEACH CLOSED

serious experimentation underway

What's going on?

Vinegar and baking soda are the two most popular kitchen chemistry ingredients on the planet. Nearly everyone knows they fizz when mixed together. Some kids even know that the bubbles are made of carbon dioxide. Few people, however, know that there are actually two chemical reactions going on. So if you want to sound really smart (and who doesn't?), then memorise the following.

In the first part of the reaction, vinegar and baking soda react to form sodium acetate and carbonic acid:

Reaction 1: Vinegar + Baking Soda →
Sodium Acetate + Carbonic Acid

The carbon dioxide gas is produced in the second reaction. It happens immediately after the first reaction because carbonic acid is very unstable. It breaks down spontaneously into water and carbon dioxide gas.

Reaction 2: Carbonic Acid → Water + Carbon Dioxide

This second reaction is the fizzing you see. The water and carbon dioxide produced weigh the same as the original carbonic acid. But carbon dioxide is a gas, so it takes up much more room. It fills all the available space and inflates the glove so it stands up straight. No ghosts – just gas.

Cooking with gas

Baking soda makes cakes and pikelets nice and spongy. Heat and reactions with acids in the mixture make carbon dioxide bubbles so the mixture expands. Bread dough rises too but via a different reaction. The carbon dioxide in bread dough is made by microscopic organisms called yeast via a process called fermentation.

35 Film Canister Rocket

Even when you know what to expect, this rocket still gives you a jolt of excitement. Film canisters are pretty light, but you still need to be careful not to get in the rocket's way! Note: the black film canister lids aren't tight enough and don't really work.

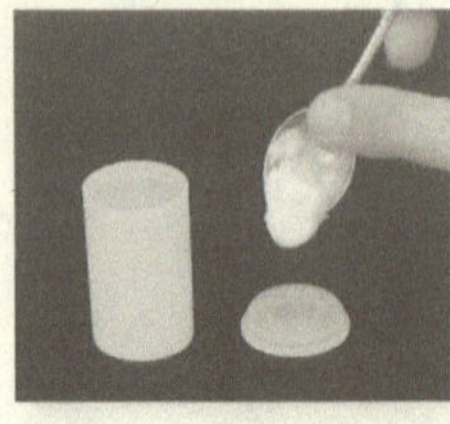

Make a thick paste by adding a few drops of water to a tablespoon of baking soda. Spoon some paste into the well inside the film canister lid. Hint: ask a photo lab if you can't find an old film canister.

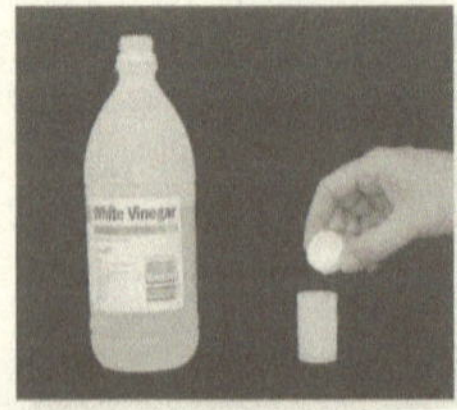

Add about a capful of white vinegar to the film canister. Carefully press the lid firmly into the canister. Nothing will happen – yet!

Pick up the can, turn it upside down, shake twice and quickly set it down. If you pick the can up as in the photo, it's easier to set it down the right way.

In a few seconds, the pressure pops the lid. The can shoots up like a rocket. This photo shows the blob of vinegar and baking soda left behind just after take-off. Cool!

What's going on?

The carbon dioxide produced by the vinegar and baking soda reaction fills the entire can. Pressure builds up and eventually, the lid shoots off with a loud pop. By turning it upside down, the force of all this pressure causes the can to fly into the air. Stick some putty to the top of your rocket for more height. Paper streamers will make it fly straighter.

Acids, bases and pH

Vinegar is a mild acid. Baking soda is a mild base. Whether a chemical is an acid or base depends on its pH. The letters pH stand for potential of hydrogen. Acids have a pH between 0 and 7. Bases are chemicals which have a pH between 7 and 14. A substance with a pH of 7 is neither an acid nor a base.

Hydrogen is the lightest element in the universe. Pure hydrogen is a highly explosive gas. But hydrogen atoms readily combine with atoms of other elements to form completely different substances. Water, for example, consists of two hydrogen (H) atoms stuck onto one oxygen (O) atom. The chemical symbol for water is H_2O. The little two means there are two hydrogen atoms in every water molecule.

The hydrogen atoms in a water molecule are stuck very tightly to the oxygen atom. That's why pure water does not react easily with other substances.

In some substances, the hydrogen atoms are not as tightly stuck. These substances can donate some of their hydrogen atoms to others and are called acids. Other substances readily accept extra hydrogen atoms. These are bases. pH is a measure of how intensely an acid or base will react with other substances. Strong acids and bases are very dangerous because they react so strongly with skin. Ouch!

36 Distraction Grenade

Every secret agent needs cool gadgets when the enemy closes in. Enter the Distraction Grenade. The loud pop distracts evil henchmen while you make a snappy escape (but only if they're a bit, er – Double-O-Dumb!)

Put a tablespoon of baking soda in a sandwich sized zip-lock bag.

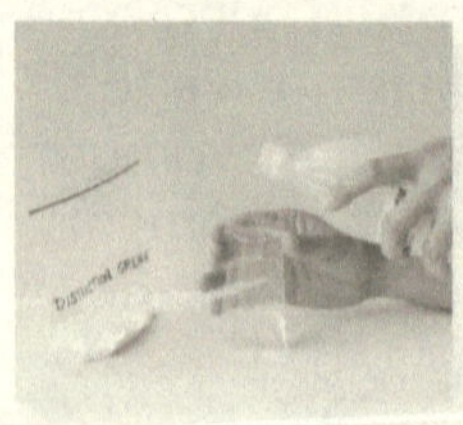

Fill a smaller zip-lock bag to the brim with white vinegar and seal it. (Hint: corner shops use small bags to sell lollies. Ask politely and the shopkeeper might give you some.)

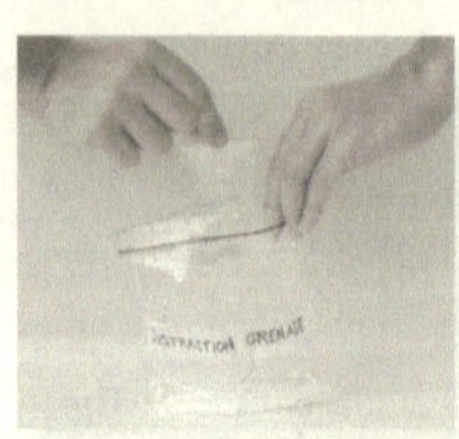

Put the smaller zip-lock bag inside the large bag with baking soda. Seal the large bag.

Do this outside on the grass or in a tub or sink. Squeeze the small bag so it pops. Throw the bag on the ground or into the sink. The bag inflates and eventually pops with a loud bang.

SCIENTISTS patrol these waters

What's going on?

The vinegar and baking soda reaction produces enough carbon dioxide to burst a zip-lock bag. The loud pop is harmless, but provides light entertainment on an otherwise stressful day. It doubles nicely as a de-stress bag.

Um...your fly's undone!

Zippers transformed the way we dress long before they appeared on plastic bags. But like many other inventions, zippers created a new problem, too. Forgetting to do them up! So let's see if history holds a solution to this terrifying and embarrassing situation.

In 1851, an American inventor called Elias Howe patented the world's first 'Automatic, Continuous Clothing Closure'. It didn't fly very well though (if you'll pardon the pun). Almost forty years later, Whitcomb Judson improved Elias's design with a scary, metal apparatus which used interlocking hooks and eyes. It pulled apart too easily so it didn't take off but the general idea was clearly a good one.

It was Swiss-born engineer Gideon Sundback who finally solved all the zip's problems. His 1913 'Hookless No. 1' wore out too quickly but he solved that problem too. The 'Hookless No. 2' quickly became a huge success. This design is still popular today! The B F Goodrich company coined the name 'zipper' in 1923. It was so catchy that 'zipper' soon became the generic name for all sliding lockers.

Now you may be wondering how this long story will help next time you forget to do up your zipper? Well, frankly, it won't. Instead, I suggest you deploy one of your new-found Distraction Grenades and then do what I do – run away!

37 Busy Water Bugs

These water bugs look like they're alive. They dive busily for food and scurry back to the surface to breathe. How? By excreting air bubbles, of course! And the vinegar smell? Oh, that's just because they go to the toilet so often – eeew!

Pour water into a glass so it's about three-quarters full. Add one tablespoon of baking soda and stir until clear.

Toss in five to ten uncooked popcorn kernels (sultanas work, too). Now pour in some vinegar until the glass is almost full.

The fizzing reaction will be fairly wild at first, but will soon settle down. When it does, the popcorn kernels start to do a groovy dance. They're alive!

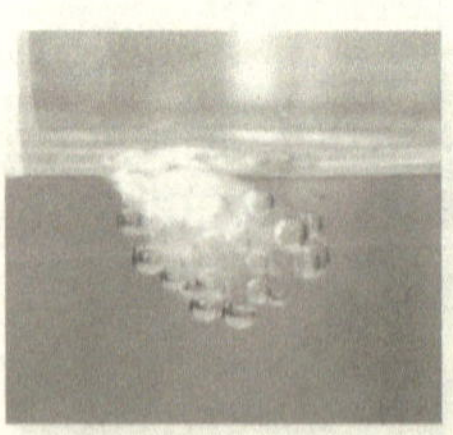

Look closely and you'll notice bubbles forming on the popcorn. At the surface, the bubbles pop and the 'bugs' may flip over. When the remaining bubbles pop, the bugs sink again.

Sandologist
ON DUTY

What's going on?

It's the vinegar and baking soda reaction once more. This time, the carbon dioxide makes your Busy Water Bugs float and sink. Uncooked popcorn and sultanas are slightly more dense than water, so they sink – but only just. The bubbles of carbon dioxide that cling to them act like tiny life jackets, so they float. These bubbles pop at the surface. If the 'bug' is completely covered in bubbles, it may do a flip before sinking again. It's a groovy little water dance!

Real busy water bugs

Visit a pond, creek or stream and you'll see heaps of critters busily making their living. Your arrival might scare them off, but sit still and they'll soon be back. The dragonflies zooming across the surface are females depositing their eggs in the water. Their nymphs will stay under water for up to two years. Then they'll climb onto a rock and moult. Adult dragonflies usually only live for a few weeks.

You might see water striders skating around on the surface, too. Air trapped between tiny hairs stops them sinking. They are hunters and scavengers. They pierce their prey with their mouths, so they can suck them dry! They escape predators by ducking under the water and resurface when it's safe.

Under the water, all kinds of critters are wriggling, swimming and beavering about. Many are the larvae of flying insects like mosquitoes and mayflies. Giant water bugs can grow big enough to eat fish, frogs and tadpoles! Freshwater streams really are amazing places if you know what to look for. So turn off the television for a while, grab your hat and sunscreen and go and visit one. You'll be amazed at what's happening down there.

38 LP UFO

Turn an old vinyl record into a flying saucer on a string. While it's spinning, it will stay perfectly flat, even while it swings from side to side. Like, whoa! What a spin-out!

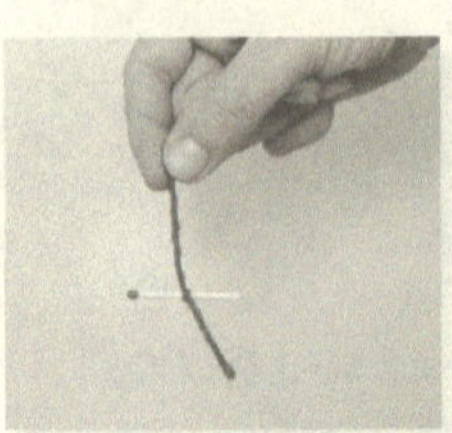

Tie a match to the end of a piece of string.

Thread the match through the hole in an old vinyl LP record. Hold onto the other end of the string.

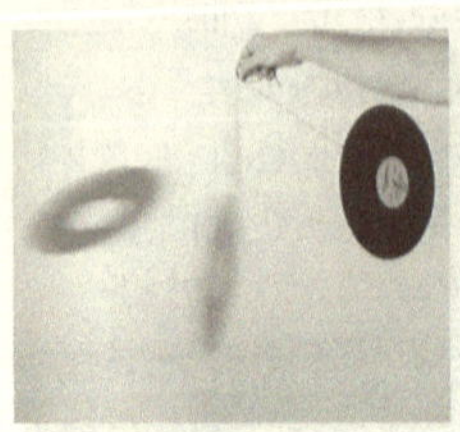

Swing the record back and forth and it just wobbles all over the place. It's a UFO, but at this point, that just means an Unstable Floppy Oscillator.

Hold the string and give the record a good spin. Get it going faster by tapping the edge. Swing the record again and this time it stays flat. Wow, an Unexpectedly Fabulous Ornament!

BEACH CLOSED
serious experimentation underway

What's going on?

You've just turned a record into a gyroscope. Like all spinning objects, the record is much more stable when it's turning. It's so stable, in fact, that it stays horizontal even when you swing it.

The reason spinning objects are so stable is their angular momentum. Momentum is what keeps an object doing the same thing. A moving train has lots of momentum, which makes it hard to stop. A bowling ball flying down the lane has lots of momentum, too, so it knocks over the pins easily. A beach ball going just as fast has much less weight and probably won't knock a single pin over. So momentum is the combined effect of speed and mass (weight).

A spinning object has angular momentum. Its mass is moving, even though the object itself is not going anywhere. It's hard to notice a spinning disk's rotation, but this angular momentum becomes very noticeable if you try to tilt it. The faster your record spins, the more angular momentum it has and the more stable it is. As it slows down, it starts to get the wobbles again. It looks a bit like the disk-shaped spinning UFOs you see in old science fiction movies!

Real gyroscopes

Gyroscopes are found in all sorts of places from toy shops to sophisticated navigation panels. They're used to stabilise moving equipment or maintain orientation (ie to remember which way is up or down). The spinning wheel disk in a gyro is fixed to a frame by a gimbal. If the gimbal is fixed, then the gyro will keep the frame stable. If the gimbal can rotate, then the disk will stay flat when the frame rotates. This tells the pilot if the plane is level, upside down or somewhere in between.

39 Bread Clip Speedboat

Float a bread clip in a plate of full cream milk. Add a drop of detergent and, believe it or not, you've got a groovy little speedboat that really moo-o-o-oves!

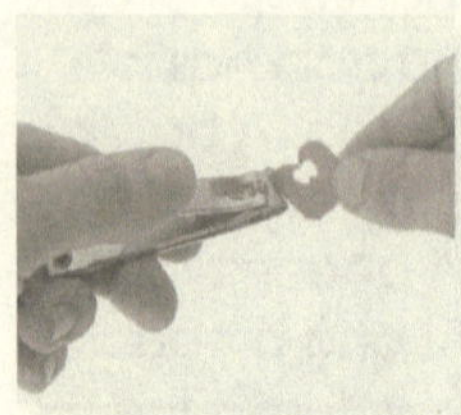

Use nail clippers to trim your bread clip into a speedboat shape.

The bread clip on the left is trimmed symmetrically. It will sail straight ahead. The one on the right will sail to the right because it is not symmetrical.

Float your bread clip in a plate of full cream milk (skim won't work). Use an eye-dropper or toothpick to add a good drop of detergent to the little hole at the back.

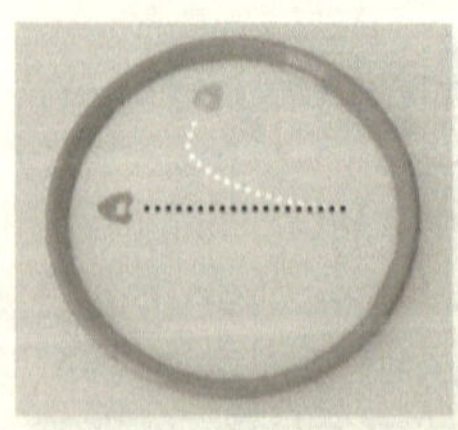

The symmetrical speedboat goes straight, but gets stuck on the opposite edge. Turn it with a clean toothpick. The asymmetrical boat goes around in circles until the detergent runs out!

SWIM BETWEEN the experiments

What's going on?

Milk is a complex mixture of tiny fat globules, proteins and sugars suspended in water. Chemists call this sort of mixture an emulsion. Fats and oils don't usually mix with water and end up in separate layers with the fat or oil on top. But the fat globules in homogenised milk are so tiny that they stay evenly dispersed.

Nobody has done an experiment to figure out exactly what detergent does to milk. Here's what we do know. Detergent has the amazing ability to mix with both oil and water. Substances that can do this are called emulsifiers. Detergent also reduces the surface tension of water. When you add detergent to the boat, the surface tension at the back is immediately reduced. Now there is more surface tension at the front. The greater surface tension tugs the little boat forward. You can see the detergent shooting out of the little hole, which also provides some propulsion. That's probably due to the difference in detergent concentration inside and outside the boat.

A symmetrical speedboat may look a bit better, but it will go straight ahead and get stuck on the edge of the plate. Asymmetrical boats putt around in a circle and keep on putt-putting until the detergent runs out. Way more interesting!

World's Fastest Speedboat

The world water speed record is a blistering 511 kmh. It was set in 1978 by Australian Ken Warby on Blowering Dam in New South Wales. He achieved his unbroken record in a jet powered boat called *Spirit of Australia*, which he designed and built himself! It was powered by a 6 000 horse power Westinghouse jet engine. *Spirit of Australia* is now permanently on display at the National Maritime Museum in Sydney.

40 Speedboat Drag Strip

Now you've mastered the Bread Clip Speedboat, it's time to go overboard. Add some food colouring to make a cool, colourful wake. If you're the competitive type, why not build a Bread Clip Speedboat Drag Strip?

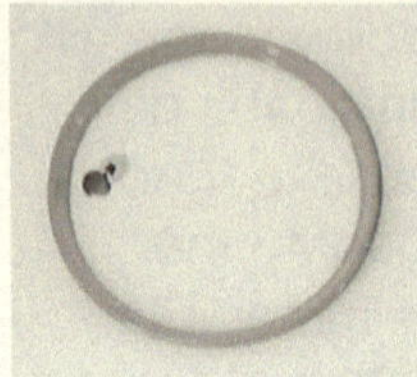

Add a drop of food colouring before you add the detergent. You'll notice that the food colouring doesn't make the boat move at all.

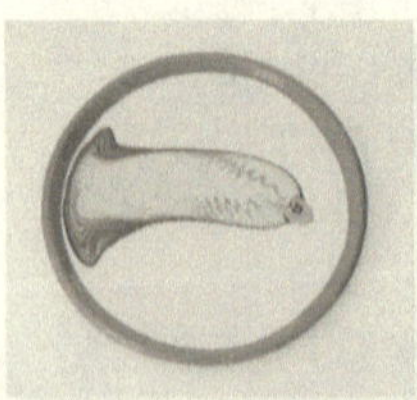

When you add the detergent, the food colouring explodes out of the bread clip. This one left a really cool double streak as it motored along.

I used the lid of a rectangular plastic tub. The lane dividers are wooden skewers. They're held down with some poster putty. I even made a finish line.

Race time! The best part is, you don't need to refresh the milk between races. It keeps working over and over again.

SCIENTISTS
patrol these waters

What's going on?

Adding food colouring doesn't affect your Bread Clip Speedboat's performance. It does make it look a whole lot cooler, though! Detergent still provides the driving force.

You can't see detergent molecules – even with the most powerful microscope. But scientists have discovered that they are long and skinny. One end, called the head, is strongly attracted to water molecules. The other end, called the tail, is strongly repelled by water. This results in some fascinating molecular behaviour. In water, detergent molecules prefer to be at the surface where they can stick their tales up into the air. When you add a drop of detergent to water, molecules zip around the surface until it is full. The remaining molecules jostle around under the surface where they latch their tales to 'hide' them from the surrounding water. There's some amazing stuff going on when you do those dishes!

Crushed beetles with that?

Food dyes are great for science experiments. Some are derived from plants, others from animals. But there's one type of red dye you'd probably prefer not to know about. It's called cochineal or carmine. It has the food additive number E 120 and it's made by crushing up cochineal beetles. Yes, beetles! The red extract from these little bugs was used to colour the coats of British soldiers on the First Fleet. In the 1920s, these little critters were used to control an Australian invasion by Argentinian prickly pears with great success. Cochineal larvae eat prickly pears! These days, cochineal is used to colour some chocolate buttons, berry flavoured yoghurts and numerous other 'treats'. Mmm...crushed beetles. Yum!

Patterns

Boy Rocking Doll **(Page 20)**

1 Photocopy or trace these dolls
2 Colour in with your favourite team's colours
3 Check which circle to cut around by placing your deodorant bottle on top
4 Cut out the doll keeping the circle attached
5 Go to page 20 and follow the instructions to attach the doll to your bottle lid

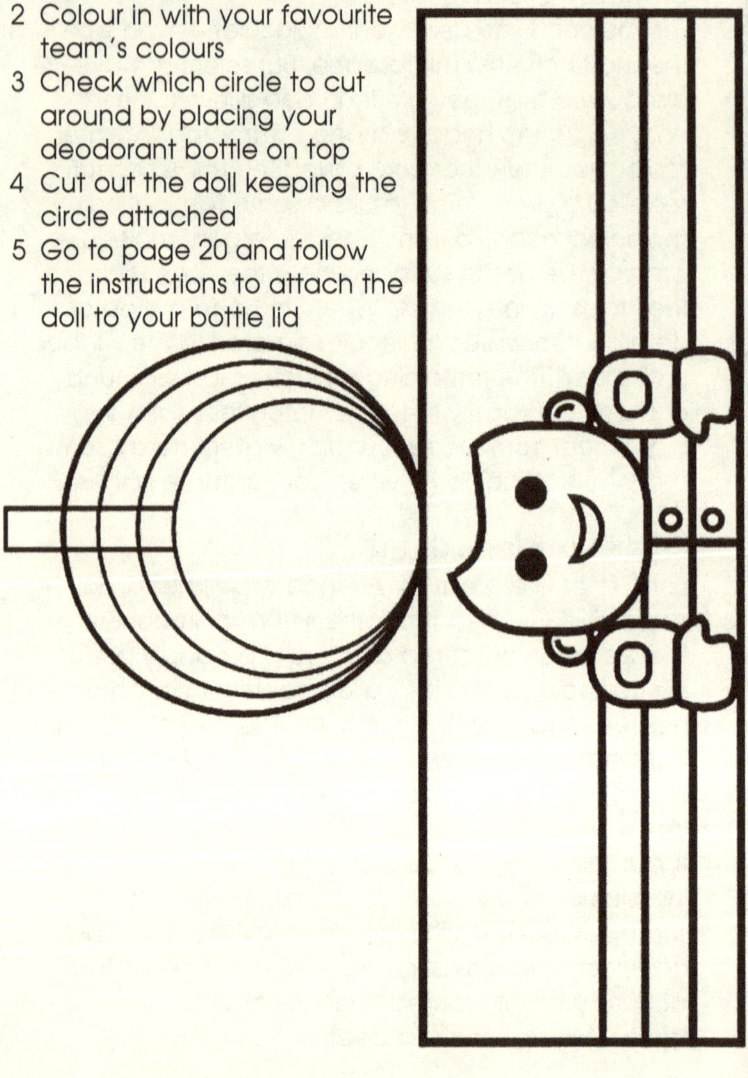

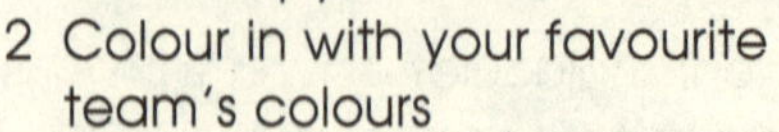

Girl Rocking Doll (Page 20)

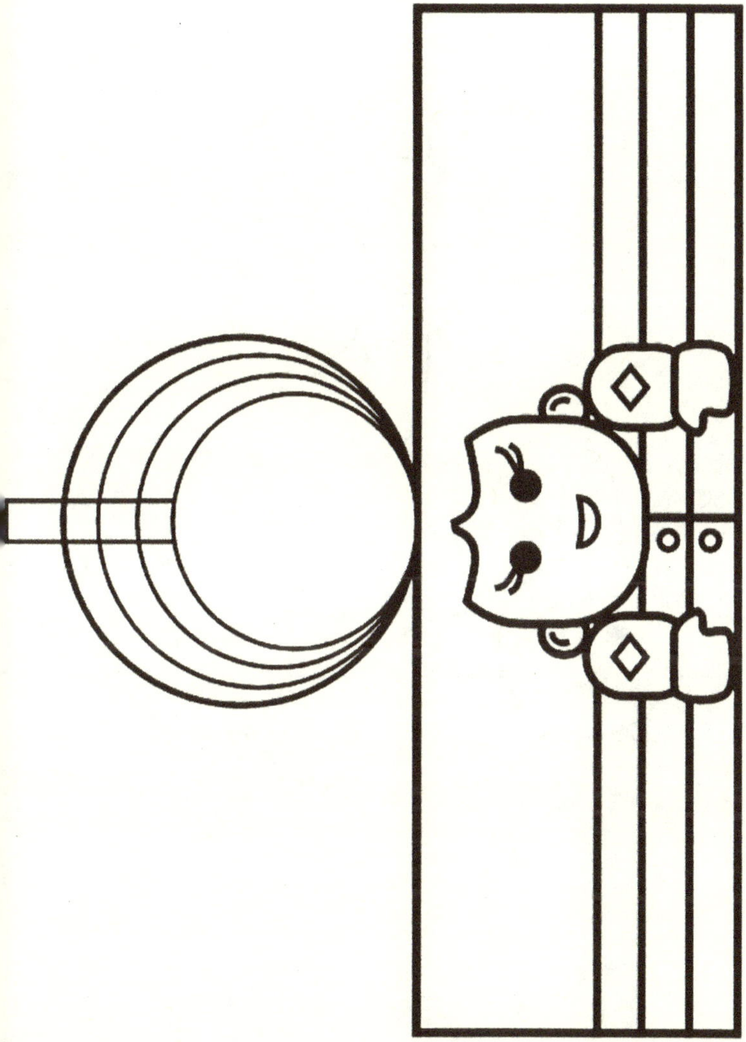

Magic Flowers (Page 32)

1 Photocopy or trace the flowers
2 Colour them in
3 Cut the flowers out
4 Follow instructions on P32

Fins for Pop-A-Rocket (Page 58)

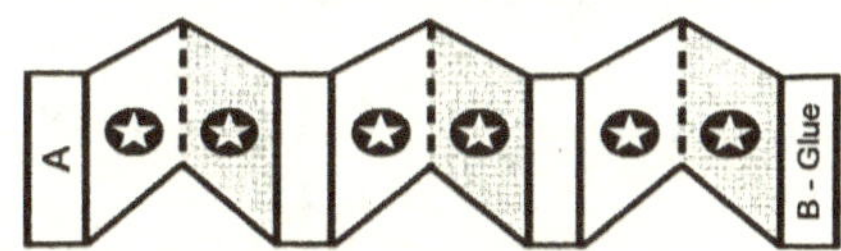

1 Photocopy or trace the designs below and cut out
2 Fold the dotted lines inwards and the solid lines out
3 Glue the back of the stars pairs press them together
4 Apply glue to the rest of the back and to flap B
5 Wrap fins around straw and stick flap A over flap B

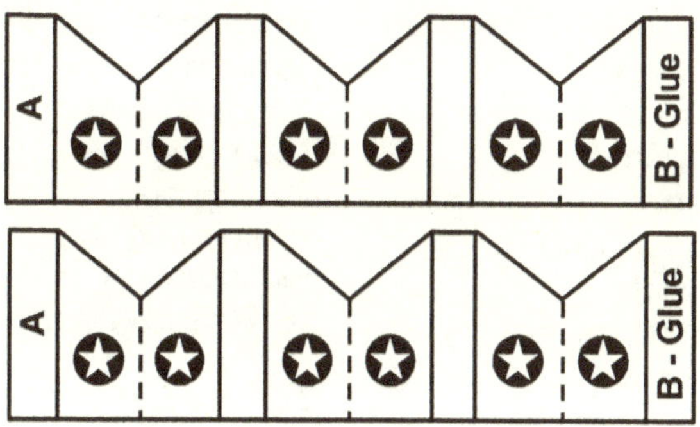

Mobile Disco Dancers (Page 68)

Photo copy and cut out the tiny dancers and fold as shown

Gizmo Equipment

Straws – bendy (various thicknesses)
Straws – clear
Sticky tape
Ruler
Scissors
Stiff card
Balloons
String
Sports drink bottle nozzle
Plastic cups
Hair dryer
Aluminium foil
Marble
Metal fork and spoon
Toothpicks
Poster hanging putty
Paper towel
Lemon
Candle
White paper
Empty soft drink bottles
Drawing pins
Vegetable Oil
Soluble tablet
CD (old)
Empty sauce squeeze bottle
Uncooked rice
Paper clips
Plastic bag
Glass
Cling wrap
Salt and pepper
Slinky spring
Rubber gloves
Food dye
Baking soda
White vinegar
Film can
Zip Lock Bag (sandwich size)
Zip Lock Bag (small)
Uncooked popcorn
Sultanas
LP Record (old)
Bread clips
Nail clippers
Full cream milk

Some Famous Inventions

It's impossible to list all the world's famous inventions. Here are just a few that really did change the world:

3500 BC *The Wheel* around 3500 BC
1200's *Glass Mirror* Unknown (Venice)
1200's *Compass* Unknown (China and/or Europe)
1565 *Pencil* Conrad Gesner (Switzerland)
1592 *Thermometer* Galileo Galilei (Italy)
1794 *Ball Bearing* Philip Vaughan (England)
1800 *Battery* Alessandro Volta (Italy)
1818 *Bicycle* Baron Karl de Drais de Sauerbrun (Germany)
1839 *Vulcanised rubber* Charles Goodyear (US)
1853 Hypodermic syringe Charles Gabriel Pravaz (France)
1859 *Internal Combustion Engine* Étienne Lenoir (France)
1867 *Dynamite* Alfred Nobel (Sweden)
1871 *The Periodic Table* Dmitry Mendeleyev (Russia)
1872 *Polyvinyl Chloride* (PVC) Eugen Baumann (Germany)
1873 *Jeans* Levi Strauss, Jacob Davis (US)
1876 *Telephone* Alexander Graham Bell (Scotland/US)
1879 *Light bulb* Thomas Edison (US)... and others!
1888 *Pneumatic Tyre* John Boyd Dunlop (UK)
1896 *Radio* Guglielmo Marconi (Italy)
1903 *Powered Aeroplane* Wilbur & Orville Wright (US)
1921 *Adhesive bandage* Earle Dickson (US)
1928 *Sliced Bread* (slicing machine) Otto Rohwedder (US)
1941 *Electric guitar* Les Paul (US)
1989 *World Wide Web* Tim Berners-Lee (UK)

Acknowledgements

I am very grateful to Ian Allen and Frankie Lee at ABC Science Online for their kind and patient support all along. To Elliot Spencer, Sally Goldrick, Emma Dobinson and the Roller Coaster team, thank you for inviting me to bring science to your astute young audience. Sincere thanks to Belinda Bolliger, Mark MacLeod and Sue-Anna Purcell at ABC Books for steering the boat, terrific editing, encouragement and support. To Sandra Nobes for the great design and cool covers. And to Kristina Schulz for talking me into writing these books in the first place!

Ruben Meerman is a surfer with a physics degree and a Grad Dip in Science Communication. He has taught primary science education at Griffith University, worked in the laser industry and performs hundreds of science shows in schools around Australia every year. He appears on ABC TV's Catalyst, Rollercoaster and The ExperiMentals and at a school near you!

www.abc.net.au/science/surfingscientist/
www.abc.net.au/experimentals

www.ingramcontent.com/pod-product-compliance
Lightning Source LLC
LaVergne TN
LVHW051015080826
845145LV00009B/2635

* 9 7 8 0 7 3 3 3 2 3 8 3 6 *